Managing Projects in Developing Countries

Managing Projects in Developing Countries

Edited by

J W Cusworth and T R Franks

Routledge
Taylor & Francis Group

LONDON AND NEW YORK

"For our wives, Jenny and Liz"

First published 1993 by Pearson Education Limited

Published 2013 by Routledge
2 Park Square, Milton Park, Abingdon, Oxon OX14 4RN
711 Third Avenue, New York, NY 10017, USA

Routledge is an imprint of the Taylor & Francis Group, an informa business

ISBN: 978-0-582-08223-6 (pbk)

British Library Cataloguing in Publication Data
A CIP record for this book is available from the British Library.

Library of Congress Cataloging-in-Publication Data
A catalog entry for this title is available from the Library of Congress.

Set by 13 in 10/11pt Palatino

Contents

Preface

This book is designed primarily for use by teachers and students of project management in developing countries and by practising managers or those about to take up managerial responsibilities. Officials in government and donor agencies involved with development projects may also find it of value. Its main focus is the establishment and commissioning phases of the project but it also includes coverage of a number of aspects of operational management.

The rationale for the book is based on two observations. The first is the increasing evidence from a variety of sources that the poor performance of project management is a major constraint to the sustained achievement of project objectives. This has led to the resultant increased demand for the strengthening of project management capacity in many countries.

The second is that, while there is a considerable body of literature in the area of project management based on the western industrial/commercial concept of a project, much of this may be inappropriate in the context of developing countries. This book attempts to discuss the concepts and techniques of project management in relation to development projects, indicating the difficulties and limitations to their usefulness in the developing economy context.

The book consists of three introductory chapters which provide a framework and perspective for the treatment of project management. Chapter 1 discusses development and the contribution of projects to development. Chapter 2 examines the project format and framework, considering in particular the interaction between projects and policy-making and the relationship between the project and the environment within which it is being implemented. Chapter 3 defines the nature of management and the role of the manager.

The core of the book (Chapters 4–10) covers the concepts, systems and skills of project management. The chapters fall into one of the two main elements of management, that is technical skills and techniques (including implementation planning, procurement and financial management) and organizational and people-related skills (in particular aspects of organizational design and human behaviour within organizations). Chapters 5 and 10, which deal with the human behaviour aspects of management, have been contributed by Dr Farhad Analoui, a lecturer at the Development and Project Planning Centre, University of Bradford (DPPC) and specialist in human resource and management development.

The last two chapters of the book serve to widen its coverage. Chapter 11 looks beyond projects and at their role in the day-to-day operations and long-term development of public agencies and enterprises. Chapter 12 examines some of the major issues in current development thinking which affect project managers. This chapter has been contributed by Dr Carolyne Dennis, also a lecturer at DPPC and a sociologist with an extensive experience of development policies and projects across a wide range of countries and organisations.

The book is a product of the combined experience of the authors, both in operational work in developing countries and in running courses in development project management at the Development and Project Planning Centre, University of Bradford. The authors wish to acknowledge the material contributions and advice they received from many colleagues at the DPPC. Inevitably some of the material drawn on to compile this book has originated from the DPPC's stock of case study and training material that several people have developed and modified over the years. Particular thanks are due to Dr Allan Low, Patrick Ryan and Dr Jyoti Majumdar, all of whom willingly offered advice and allowed the authors to adapt some of their own material for use in the text.

The authors also wish to acknowledge the hard work and effort made by Mrs Jean Hill and Mrs Judith Jaques in typing and retyping the drafts of the text.

The Authors

Tom Franks started his professional life as a civil engineering contractor in the UK and Saudi Arabia. He then spent ten years as an irrigation engineer with a consulting firm, working on a variety of water resource projects in north-east Africa, Pakistan and Indonesia. In 1984 he joined the Development and Project Planning Centre, where he has been responsible for developing the DPPC's training courses in the management of capital projects. He has continued his overseas involvement with a number of consultancy and training assignments in project management, and has also been responsible for co-ordinating the DPPC's institutional development programme with the Ministry of Finance and Planning in Sri Lanka. Since 1990 he has been on a major Anglo-Japanese – supported flood control project in Bangladesh.

John Cusworth is an agricultural economist. After three years working for an Oxfam–supported rural development project in North Western Province, Zambia, between 1971 and 1974 he spent five years working for the UK Overseas Development Administration (ODA) in operational postings in the West Indies and Sierra Leone. In 1979 he joined the Project Planning Centre at Bradford and directed the annual Planning and Appraisal of Rural Development courses there from 1980 to 1983. In 1984 he took long-term leave of absence from the DPPC and spent two and half years in Zimbabwe working on the ODA–supported Land Resettlement Programme. Since returning to Bradford his main teaching role has been as director of the now well-established course in Agricultural and Rural Project Management. He is also currently the director of one of the DPPC's three programmes, responsible for developing the consultancy and commissioned training activites of the DPPC.

Farhad Analoui is a psychologist with a research background in human behaviour in the workplace. He specializes in management development and has been responsible for developing a new post-experience course at the DPPC in Human Resource Management for Development Project Managers as well as undertaking several training needs related assignments overseas on behalf of the Centre.

Carolyne Dennis is a sociologist with extensive experience of developing countries, including fifteen years undertaking teaching and research at the University of Ife, Nigeria, between 1970 and 1984. Since joining the DPPC she has undertaken a wide variety of assignments overseas on behalf of the DPPC and other agencies. She has also developed and directed the post-experience course on the Planning and Appraisal of Rural Development Projects and co-ordinated the DPPC's Masters Degree programme in National Development and Project Planning.

CHAPTER 1

Development and development projects

The policy framework

This book is concerned with the practical management of development projects. Development is used here in the sense that some countries are described as being 'developed' while others are 'developing' or 'under-developed'. It is a concept which has become widely used only since the Second World War. A variety of theories have been advocated as the best method of achieving development. The neo-classical approach emphasized national economic growth based on investment and growth theories such as the Harrod Domar model (Sen 1970) in which particular emphasis was put on industrial expansion. The approach was dominant in the 1950s and 1960s, both in the free market and centrally planned economies. A development of the neo-classical approach was the 'trickle-down' theory, which suggested that all members of society would benefit from national growth, as increased wealth gradually spread from the richer sections of the community to the poorer. When this appeared not to be happening, the neo-classical approach based on industrial growth was replaced by an emphasis on the direct satisfaction of basic human needs (food, shelter, health, transport and education), particularly for the poorer members of society. The concept of development then began to assume a precise form as first, the satisfaction of basic human needs and, beyond that, as giving people the capacity to determine their own future. Throughout this period projects played a key role, because they seemed to represent the most practical method of achieving specific goals and targets in both the neo-classical and basic needs approaches, and were a way of concentrating and combining scarce human and material resources to achieve maximum effect. Indeed they were particularly appropriate to the neo-classical approach with its emphasis on the expansion of production; for a considerable time the word 'project' came to be associated almost exclusively with the construction of industrial, infrastructural or directly productive facilities.

The past few years have seen a shift in emphasis in development. In the 1970s and early 1980s the focus was on the study and improvement of projects as a mechanism of successful development, whether directed

1

towards growth or the satisfaction of basic needs. In the late 1980s and early 1990s the emphasis has shifted to the study and analysis of policies, focusing on the general direction and framework of government measures, rather than specific actions represented by projects. This shift has been assisted by the increasing importance being given by the international lending agencies to balance of payments support and structural adjustment loans, and the corresponding decrease in the importance of project lending. At the same time attention is being paid to the potential of private enterprise to provide mechanisms for development, and there is a growing awareness of the need to enhance the efficiency of organizations, particularly in the public sector, through processes of institutional development.

Policies determine the environment and framework within which development takes place. Get the policies right, it is argued, and successful development will follow. Nevertheless, the tactical processes of development also need attention and, for the foreseeable future, projects are likely to form a major part of these tactics. Projects and the project approach are an instrument of policy, and are one means by which policies are put into practice. The change, which is inherent in any form of economic, institutional or social development, is brought about by initiative, impetus and, where necessary, capital investment, which may be provided by a project.

The need to link appropriate policies to appropriate projects is an increasingly important element of the development process. Whatever their shortcomings, projects will remain as an important mechanism for implementing policies: they are, and will remain, demonstrations of the effects of policies at a practical level. They also provide a means of assessing the impact of development initiatives on people. For example, a policy to attain self-sufficiency in rice may well be implemented through a series of projects related to the supply of irrigation facilities, development of improved seed, and provision of related inputs, such as training and marketing. A review of these projects, together or singly, increases our knowledge of the possible effect of the policy both on the economy and on individuals such as the farmer and the consumer. Similarly, a structural adjustment programme may include a requirement to reduce the size of the public sector. The necessary retrenchment will be affected by a series of project-type initiatives, whose impacts on identifiable individuals may be readily recognized.

Although projects in general will remain important tactical development tools, different types of projects are emerging as policy frameworks change. Hitherto, a development project has tended to mean an externally funded initiative undertaken by the public sector, generally resulting in the creation of physical assets. It is inevitable that many projects are conceived and implemented in the public sector because of the relatively large size of this sector in developing countries. Increasingly, however, projects will be internally conceived and funded initiatives, undertaken by both the public and the private sector, and often concerned as much with skill enhancement and institutional development as the creation of physical assets. Whereas the typical project of the early 1970s was the development of a sugar estate and the construction of a sugar factory, the typical project of the 1990s is management development for staff in public enterprises, or review and improvement of systems of cost recovery in water undertakings. Such projects require little, if any, physical construction work, but they are nevertheless real projects, requiring similar attention to their planning and management to be successful.

What is a project?

While the meaning and theories of development provide a general background for this book, a thorough understanding of projects is fundamental to consideration of project management. Successful project managers must fully understand the nature of projects and how these differ in many important respects from other activities in which they might become involved. Indeed, many of the problems of project development stem from an inability on the part of those involved to grasp the intrinsic differences between, for instance, a project to develop a primary health care service in a rural area, and the subsequent operation of that system.

There have been many attempts to define projects, because of their important role in the development process since the 1950s. In this context, it is valuable to use a definition which is as simple and generally relevant as possible:

A project is the investment of capital in a time-bound intervention to create productive assets.

The energy and inventiveness of people play a role in projects which is as important as the expenditure of physical and financial resources, so that in this definition 'capital' refers as much to human as to physical resources. Similarly, the assets created may be human, institutional or physical. This definition of a project allows us to use it across a wide spectrum of human activity. Individuals can undertake personal projects: for instance, learning a language in order to be able to use it in business or enjoy its literature is just as much a project for that individual as building a house, though the time-scale and scope of financial investment may be vastly different. More usually, we are concerned with projects undertaken by groups of individuals and society as a whole through government. In this case projects can cover a whole variety of initiatives; these may range from those designed to enhance potential in specific groups, perhaps creating small-scale enterprises for the rural poor, through projects intended to establish new organizational forms and sets of procedures, for instance for delivering health care more efficiently, to projects for the construction of physical assets such as factories. The key aspect that distinguishes a project from other forms of investment, whether for society or an individual, is that the investment is outside the scope of the normal day-to-day or year-to-year expenditure and effort, that it takes

3

place over a particular time (in other words it is 'time-bound') and that it is intended to achieve a specific objective or set of objectives.

In arriving at a precise understanding of the nature of projects, it is also necessary to be aware of the link and distinction between projects and programmes. A programme resembles a project in that it is a set of activities designed to facilitate the achievement of specific objectives but generally on a larger scale and over a longer time frame. Characteristically the activities of a programme may be diverse in scope, and widely diffused, both in space and time. Examples are found right across the development spectrum. In response to the International Drinking Water Supply and Sanitation Decade (the objective of which was to provide all people with safe water and sanitation facilities) many countries formulated programmes consisting of a series of activities such as surveys of existing facilities and resource potential, procurement of necessary equipment for drilling and construction, widespread health education, training of skilled personnel, and so on. A programme in the industrial sector might be the planned expansion of plant capacity for sulphuric acid production in order to achieve the goal of making a country self-sufficient in a given time. Programmes may also mean much smaller routine or repetitive activities falling within an overall plan, such as the construction of grain stores or the allocation of money and personnel to an Irrigation Department for the rehabilitation of small irrigation schemes.

Development projects are often the constituent activities of programmes. In the case of water supply, for instance, the construction of a well for a village community would constitute a project, as would the construction of a dam and pipeline for an urban supply. In the case of industrial production, the construction of a new factory would be a project. The distinction between projects and programmes is not always clear-cut since many characteristics are common to both activities. A project large enough in time, scope or cost may often be called a programme; integrated agricultural development programmes/projects are a case in point. Generally, however, the important distinction is that programmes are diverse sets of activities over a long period designed to attain certain objectives, while projects have a defined starting and finishing time. Projects also tend to be location-specific, though this is not invariably the case. In this book project management relates both to individual investments (projects) and programmes.

The important characteristics of projects are that they involve capital investment (through the incurring of costs) over a limited time-frame. Projects create, over that period, assets, systems, schemes or institutions, which continue in operation and yield a flow of benefits after the project has been completed. Once an individual has expended hard work, and perhaps incurred financial costs, to learn a foreign language (the project), he or she can continue to use and enjoy that knowledge. Once a water supply project has been completed (through the commitment of investment resources such as money, skills, equipment and materials), it creates a system which is operated to supply drinking water to its consumers on a continuing basis. Unfortunately development practitioners often blur the distinction between projects and the assets, systems, schemes or institutions that they create. This decreases the effectiveness of development through the project approach, because the techniques and approaches appropriate to the time-bound investment of a project are not necessarily appropriate to the continuing operation of the assets. Another problem, which frequently occurs, particularly when large quantitites of aid funds are involved, is to view them as self-contained initiatives, separate and distinct from the other activities of the organization which owns them. In fact, any organization is likely to be involved with a number of projects, as well as a collection of ongoing activities or a portfolio of continuing business. Very often there will need to be a trade-off between projects and the other activities, perhaps in relation to the commitment of scarce resources. Project managers need to be aware of the wider framework within which they are acting, and to understand that the best interests of the organization as a whole may sometimes be better served by putting first the requirements of the operation of existing assets and systems at the expense of project development of new assets. This is often made difficult because (as will be discussed on p. 8) current practices of aid funding tend to emphasize capital investment for projects, at the expense of recurrent funding for operations.

The project cycle: the traditional approach

The idea of development projects as the time-bound creation of physical assets led in turn to the recognition of phases within the project process and from there to the concept of the project cycle. The following section discusses fundamental modifications to the project concept which require a major reassessment of this general approach. Nevertheless, the idea of the project cycle still has much of value for project managers, and serves as a useful basis for understanding. Many versions of the project cycle have been produced, all of them having as their basis the idea that projects go through a number of clearly defined stages in the process of their establishment. A well-known and influential version of the cycle presented in a cyclical form is that due to Baum (1978), while UNIDO (1979) have produced a linear form. Figure 1.1 presents a modified version which borrows something from both these models.

Whatever their differences, most models of the project cycle have the same basic concepts and highlight the following important stages: identification, formulation, implementation, commissioning, operation and evaluation.

Identification

Identification is the stage at which the project is defined as an idea or possibility worthy of further investigation and study. This may come about either as the result of the discovery of a resource which could be exploited (a valuable mineral deposit in a remote region) or a need or demand to be satisfied (inadequate skills in a particular group of agricultural extension workers). Of course, many more projects are identified than actually pass through the remainder of the cycle to completion and operation.

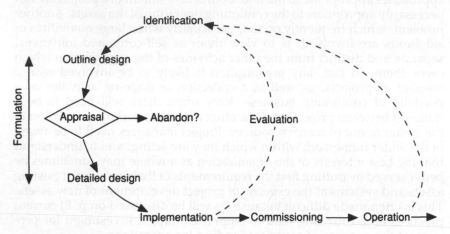

Fig. 1.1 The project cycle

Formulation

The formulation (or preparation) stage involves the definition of alternatives for the project, followed by the selection and planning of the optimum alternative, covering such aspects as size, location, technical details, markets and institutional arrangements. Thus, for a mining project, preparation would involve defining and costing the techniques and facilities required for mineral extraction and assessing the potential market and revenue, together with important environmental and institutional considerations. For the strengthening of an agricultural extension effort it would involve identification of the skills which were lacking, assessing methods of providing those skills, perhaps through a particular type of training programme, and estimating the benefits expected to accrue. Within the formulation phase certain clearly defined stages can normally be distinguished: outline design, appraisal and detailed design.

Outline design

This is the design process carried to a sufficient level of detail to allow the estimation of technical, social and institutional parameters, and the preparation of a feasibility study with an assessment of costs and benefits. This should be done ideally to accuracy of, say, 20 per cent, though lack of good data and difficulties of forecasting mean that such accuracy is often not achieved.

Appraisal

Appraisal is the process in which all aspects of the project are reviewed, in order that the decision whether or not to proceed can be made. Appraisal should cover technical, financial, economic, social and organizational aspects of the project; others, such as environmental, administrative, gender or political impacts, may also need to be considered. Where aid funds are involved, appraisal is often seen as a formal process and a clearly definable event in the project calendar which is accompanied by major negotiations between the lending agency and the host government. In reality, the decision-making implied by appraisal should go on throughout the preparation phase, as those working on the project continually review and refine the proposals. Such a process, even if it is not formally defined as appraisal, should certainly not be considered as being restricted only to aid-funded projects.

Detailed design

Following appraisal of the project plans, and a decision to proceed, designs are carried out to a sufficient detail for implementation to go ahead. These designs will include specific organizational forms and institutional procedures, and construction documents for the required physical facilities and buildings, with costs estimated to within, say, 10 per cent. Again, such accuracy is desirable, but may often not be achieved.

Implementation

Following detailed design there may well be another period of appraisal and negotiation to finalise details of the plan. When this has been completed, implementation of the project can commence. Implementation is the stage at which the institutions are established and facilities constructed. It is the stage which involves the disbursement of the largest portion of the project funds.

Commissioning

Between implementation and operations it is possible to distinguish the process of commissioning or mobilization when the constructed systems or assets are first put into operation. This process, though ignored in most versions of the project cycle, is important to subsequent successful operation of the assets and will be covered fully in Chapter 11.

Operation

The operational phase of the project is the period during which the assets created by project implementation are put to work and yield a flow of benefits (the mine is producing minerals which can be processed and marketed, or the extension workers are trained and working with farmers). Strictly speaking, the project phase has been completed by the time operations commence.

Evaluation

Evaluation consists of investigating and reviewing the effects of the completed project, to see whether the benefits which were planned to flow from it have indeed been realized, and whether these benefits have had their intended consequences.

The model of the project cycle shown in Figure 1.1 differs from earlier versions in several important ways. First, the cyclical form retains the idea that project development should be a learning process, with the evaluation of completed projects feeding back into new and improved projects in the future (though it is open to question how often this occurs in practice). Second, early versions of the cycle highlighted the various stages related to planning and implementation but often failed entirely to include the operational stage. This is the stage when the constructed or established facilities are put to use and yield benefits. It is the reason for which the project was implemented in the first place but its omission from the cycle reduces its importance to the point at which the project phase becomes an objective in itself. This is partly a reflection of the dominance of economic planners in governments and the organizations of the major international lending agencies: the undoubted importance of rigorous planning and appraisal was aided by an intellectual interest in these phases which was not matched by an equal interest in the subsequent, and equally important, phases of implementation and operation.

Another of the unfortunate consequences of an approach which overemphazises planning and implementation is that projects remain projects long after they should be considered as being operational assets. Too often, managers of projects in developing countries know that the key to their continued access to funds is to ensure that their activities remain as projects, which are financed through the capital investment budget. By contrast, facilities which are already established and in operation (project phase completed) are financed through the recurrent budget. The capital investment budget is often made up partly of counterpart foreign funds through loan agreements and partly of local funds which must be provided as a condition of the loan. Recurrent budgets, on the other hand, are generally made up entirely of local funds and are, of course, the first to be cut back at times of financial stringency, with consequent detriment to the effective operation of completed project assets.

The project concept: blueprint or adaptation

The general concept of the project taken up within the notion of the project cycle has sometimes been known as the blueprint approach. This uses the imagery of blueprints of engineering drawings to suggest that projects need to be systematically and carefully planned in advance, and implemented strictly according to the defined plan. It has, in general, proved itself to be a useful approach to investment in 'capital-intensive' projects. These are projects in which relatively large amounts of resources are expended in the implementation stage and which normally result in major physical assets, for instance the construction of industrial plants and infrastructural facilities. In these projects the completion of the construction generally marks a clear end to the implementation phase.

There is, however, another important class of projects which are often 'people-based' occurring mainly in the agricultural, rural and social sectors. These involve little in the way of financial investment, but emphasize human or institutional development, such as the development of health care training of medical extension workers. (Another distinction is between 'physical' projects, which equate to the concept of 'capital-intensive', and 'human-orientated' projects, which are similar to 'people-based' projects.) In people-based projects the blueprint approach has needed modification for a number of reasons. First, the distinction between implementation and operations in these sectors is often not clearly defined. A project to improve agricultural extension services, for instance, may well encompass training of extension agents and construction of centres in one area (implementation phase) while extension activities (operations phase) are already taking place in another area, though both sets of activities are still under the same administrative structure.

A more fundamental critique of the blueprint approach, however, emerged during the late 1970s and early 1980s. This position suggested that the blueprint approach was too rigid and inflexible, that it placed too much reliance on prior comprehensive data-gathering, planning and control (all of which often appeared to be inadequate in developing countries), and that it did not give sufficient importance to the acceptability of the proposed intervention to the intended beneficiaries. Out of this position grew the idea of a development project as an adaptive approach (Rondinelli 1983), with successive stages of experimentation, pilot, demonstration and replication or production, as shown in Figure 1.2.

In this approach, experimentation is the stage at which development problems or objectives are defined and possible solutions, methods of

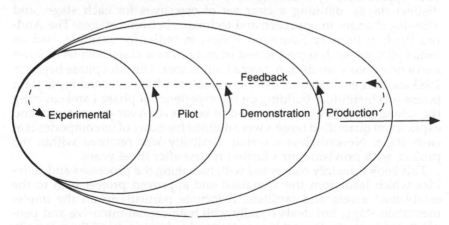

Fig. 1.2 Projects: the adaptive approach

9

analysis and implementation are identified. These are then subject to pilot trials for their appropriateness, adaptability, acceptability and transferability to local conditions. As particular methods are found to be successful on an experimental scale, they are taken successively through wider and more comprehensive coverage in order to reach a larger public, each stage growing out of the results and methods of the previous one. Intrinsic to the adaptive model is the idea that, while a general objective and direction for project intervention can be defined, specific actions and services may change as a result of review and evaluation of individual phases. This is sometimes called the 'process' approach, to highlight a flexible two-way interaction between project agents and the consumers or beneficiaries for whom the project is intended. It might result, for instance, in redesign of systems for delivering primary health care or relocation of health centres, following earlier design and experimentation. It does not change the basic concept of a project as a time-bound investment to achieve specific aims. The need or opportunity for such investment must still be identified, the investment must still be prepared and implemented (perhaps with successive stages of experimentation and modification if an adaptive approach is followed), and then the systems or facilities operated to create benefits.

The general desirability of an adaptive, flexible approach to project development is clear. It has found a more ready application in the rural and social sectors than the industrial and infrastructural sectors, partly because rural and social development is intimately affected by the reactions of people (as project agents and beneficiaries) and is therefore less easy to plan and define in advance. By contrast the blueprint approach remains important for capital-intensive projects where there is a paramount need to plan and account carefully for expenditure incurred in the public sector. This constraint allows comparatively little room for the flexibility and adaptation provided by the adaptive approach.

A further, more recent development has been to combine elements of both approaches, by breaking the project idea down into discrete and distinct stages, defining a clear set of objectives for each stage, and allowing changes in approach and technique between stages. The Andhra Pradesh Primary Education Project in India, for example, had an initial pilot phase designed to test training in new classroom techniques and a new classroom design, over a limited area. The pilot phase began in 1983 and was evaluated in 1987, as a result of which a second replication phase was instituted, building on the experience of phase I and covering the whole state. This was expected to be a seven-year programme, and explicit and quantified targets were defined for many of its components at each stage. Nevertheless a certain flexibility was retained within the project, with provision for a further review after three years.

This book is mainly concerned with managing the processes and activities which lead from the appraised and approved project plan to the established assets and facilities. It focuses particularly on the implementation stage, but deals equally with both capital-intensive and people-based projects. Project managers need to keep in mind the necessity of managing and implementing the project as planned, while at the same time retaining as much flexibility as possible in order to make changes to the overall design where necessary. Another necessity for project managers, whether the blueprint or adaptive approach is followed, is to see beyond the project phase and to ensure that the efforts of their project team lead to the creation of lasting and beneficial assets.

Why projects succeed or fail

Part of the reason for the increasing emphasis on policies at the expense of projects is the perception that many projects have failed to achieve the targets set for them, or that they have unintended negative consequences which far outweigh their direct benefits. Much can be learnt by looking at the reasons for such failure. Indeed, it is a good deal easier to do this than to cite the reasons for a successful project, partly because it can often seem that a single overriding factor is responsible for failure (though this is frequently an erroneous assumption), whereas success must be the result of several factors working together and it is hard to distinguish the most significant.

Project failure can in fact be identified at two levels. First, there is the failure to implement the project effectively, that is on time, within budgets and according to the plan (which may itself have been revised during the course of implementation). It is a common occurrence to find projects running late, and therefore probably suffering cost over-runs due to inflation. For instance, the World Bank's annual audit of projects regularly finds up to 30 per cent of the projects surveyed more than six months late. Second, more obvious failures take place, however, when implementation has been completed but the facilities created fail to achieve the effects intended. Examples of such failures abound: the water points that are unused because they are wrongly sited or which cause desertification due to over-use, and the meat processing plant that operates at half capacity because local people are not willing to sell their cattle in the expected numbers. Project managers will be concerned to avoid both types of failure, as far as it lies within their power, but they will be more directly involved with failure to implement the project effectively. Responsibility for failure to achieve the effects intended lies more often with the planners and designers, rather than the managers.

In fact, a useful approach when studying project failures is to distinguish between those situations where the causes of failure are internal to the project process (and therefore susceptible to some degree to influence and control by the planner and manager) and those where they are external to the project and therefore not susceptible to influence in the same way. Included in the former category are such factors as faulty assessment of the resource base (for instance, insufficient suitable land for an agricultural project), poor planning, problems with procurement and technology transfer, lack of resources such as finance and skilled personnel, organizational inefficiencies and the like. Even some of these factors may not be completely avoidable through greater efforts by project staff. Professionals in the development field typically show a belief in the efficacy of systematic and comprehensive planning and management which is not borne out by experience. Development is a complex undertaking and it is often impossible to take all the factors into account in the planning process.

External factors contributing to project failure are the subject of much discussion and debate. These are often referred to as project environmental factors, a subject covered more fully in Chapter 2. The project environment means much more than simply the physical or natural environment, though that is an important part of it. It means the whole set of people, things and institutions which surround the project and interact with it. At one level a set of environmental factors can be identified, such as natural disasters (for example, a flood) or external economic influences (for example, a devaluation of currency) which are not, except in the broadest sense, predictable over the lifetime of a project. Perhaps a more important set of external factors are those caused by the reactions of people affected by the project. These may show at a variety of levels. A common reason given for failure of a project is lack of high-level political commitment to it, and the lack of strong management and leadership that often accompanies this. Another common reason is the cultural misfit of the project's objectives and activities within the environment and a lack of local knowledge and understanding, leading to rejection of the project by the intended beneficiaries. For instance there have been some notable failures in agricultural projects in West Africa because planners assumed that men would own and farm the land whereas in fact women play a pivotal role in agriculture in this region. The value of the adaptive approach to project development lies in its ability to take into account factors which were not known at the time the plan was made and to be flexible in response to them.

Project formulation itself is, therefore, beginning to adapt to the need to fit into the local situation and to develop an approach which is in harmony with existing practices. In rural development, for example, the days of large area-based projects requiring massive investment in infrastructure have been superseded by projects aiming to achieve the same objectives through support of existing systems. The case of the Sierra Leone Integrated Agricultural Development Projects (IADPs) illustrates this.

In the early 1970s the Sierra Leone Ministry of Natural Resources initiated a number of these projects with the support of the World Bank and the European Community. By the end of the decade much of the country was covered by such projects, each established along similar lines. They developed management structures which employed personnel from various other institutions such as ministry departments, research stations and the university. The project managers themselves were invariably expatriates employed by the donor agencies.

The objectives of the projects were to stimulate agricultural production, increase rural incomes and improve rural welfare standards. The approach involved the building of rural infrastructure such as feeder roads, water supplies and farm service centres and the establishment of credit, input supply and extension services.

During implementation these projects, which were autonomous units, rarely worked with or through any existing institutions. The project managers were judged almost exclusively on their ability to achieve measurable project outputs such as kilometres of roads built, service centres erected, or wells dug. Far less attention was paid to building up the institutional framework for sustained development. In fact the very existence of the IADPs was steadily destroying the institutional framework of the Agriculture Department, and other departments, as more and more of their trained and experienced staff left to join the projects, which were in fierce competition to employ the more able managers, as reflected by the greatly enhanced terms of employment. This had grave consequences for continued development once the external funding for the projects was exhausted (Cusworth 1983).

A gradual realization of the potential difficulties of sustaining operations following the establishment phase led eventually to a change of approach in the Bo-Pujehun IADP in South East Sierra Leone, which was established in the early 1980s. This was another area-based externally funded project but, instead of creating an autonomous project unit, the management team simply facilitated the achievement of project objectives by making resources available to existing local organizations which managed the implementation of the various activities themselves, albeit with some assistance.

The requirement for management in this project differed substantially from the earlier projects. The more aggressive target-orientated managerial approach involving almost total control over autonomous project activities gave way to an approach requiring a greater emphasis on stimulating community participation and strengthening the capacity of the local organizations.

This approach to rural development projects in Sierra Leone was adopted because it was considered more appropriate for achieving sustained development. It built on existing strengths and reduced duplication of effort, which was a severe criticism made of the earlier approach. It is also a more culturally acceptable approach in that, by working through local organizations, the management team could be sensitive to local factors. The development of indigenous management systems which are an integral part of local culture and understanding is now an area of growing importance.

Just as lessons can be learned from the relative failure of the early rural development projects, so also some interesting research has recently been concerned to identify causes of success in project development. Paul (1983) studied six large programmes in widely diverse cultural and institutional settings. He concluded that the most important factors contributing to success are a supportive environment, a flexible, realistic project strategy and set of objectives (in other words, an adaptive approach), an appropriate organizational structure and effective management. (In this study Paul noted participation in decision-making, motivation, staff development and monitoring as key aspects of effective management.) A similar study carried out in the United Kingdom (Morris and Hough, 1986) was concerned to identify preconditions of success in major projects which were all highly capital-intensive and therefore required comprehensive planning and control. While many of the factors identified are related to the need for efficiency in these processes, the study also highlighted the need for a clear analysis of environmental factors, a supportive political climate, appropriate organizational structures, firm leadership and good communications.

While useful lessons can be learnt from these and other studies of project failure and success, project managers must be aware that each project in its environment is unique, that no two sets of circumstances will exactly replicate each other, and that it is impossible to rely solely on past experience to be a successful manager.

References

Baum W C 1978 The World Bank project cycle. *Finance and Development* **15** (4): 10-17.

Cusworth J W 1983 Integrated agricultural development projects in Sierra Leone: some implications for the future administration of agricultural development. *Journal of Agricultural Administration* **18**: 61-85.

Morris P W G, G H Hough 1986 *Pre-conditions of success and failure in major projects.* Chichester, Major Projects Association.

Paul S 1983 *Managing development programmes: the lessons of success.* Boulder, Colorado, Westview.

Rondinelli D 1983 *Development projects as policy experiments.* London, Methuen.

Sen A 1970 *Growth economies: selected readings.* Harmondsworth, Penguin.

UNIDO 1979 *Manual for the preparation of industrial feasibility studies.* Vienna, UNIDO.

CHAPTER 2

The project framework and the project environment

Introduction

Before examining the concepts and practices of project management, which are the main themes of this book, it is useful to discuss two aspects of project development which can be of value in giving project managers a clearer understanding of the nature of projects and the activities with which they are involved. These aspects are the project framework and the project environment. The project framework helps to clarify and conceptualize the project and is a useful tool upon which project managers can build up their understanding of it. The project environment is the surroundings and situation within which the project is implemented. Chapter 1 discussed reasons for project success and failure and distinguished between internal and external causes. The concept of the project environment provides managers with a structured approach to analysing and handling possible external constraints.

The project framework

The 'project framework' (or 'logical framework' as it is sometimes known) was initially developed primarily as a tool to assist with the design, preparation and evaluation of projects. Many donor agencies have since adopted variants of the framework as a basis for planning, appraising and subsequent evaluation of the projects that they support. Project managers, though they have a somewhat different responsibility towards the project, are able to use the idea of the project framework equally as much as donor agencies. It serves as an aid to help them clarify the project objectives, as a tool to increase their understanding of its various components and as a basis for establishing a monitoring system.

The project framework defines clearly the project objectives in such a way that logical linkages are established between a hierarchical set of sub-objectives, each set of which ultimately contributes to the final developmental aim of the project. The objectives are described in a vertical column showing project inputs being transformed into project outputs, which contribute to project purposes and ultimately project goals. This column, called the 'narrative summary', is linked to a second column which defines the main assumptions on which the linkages between inputs, outputs, purposes and goals are based. The third column identifies quantitative indicators which measure the achievement of the objectives of the project at its various levels. A final column defines the means of measuring such project indicators. The framework is thus as a matrix which has both a vertical and horizontal logic. A simplified project framework relating to a rural water supply project is shown in Table 2.1 and is discussed below. Readers may like to follow this discussion by drawing up a project framework for their own project.

Narrative summary

The main objectives of the project in the narrative summary context are defined as project goals, purposes, outputs, and inputs. Goals define the broad development objectives which, for most projects, tend to be described in economic or demographic terms either at the national or regional level. In our example the project goal is an improvement in health in the rural population. Examples of goals for other projects might be an improvement in the balance of trade or in rural standards of living. Project purposes are generally described in terms of the factors that contribute to the achievement of the overall objectives, for example increased consumption of safe water, production of goods for export or improvement in the levels of nutrition among poor families. (In some versions of the project framework 'goals' and 'purposes' are called respectively 'impacts' and 'effects'.)

Project outputs contribute to the achievement of project purposes. They may not be sufficient by themselves to achieve these purposes: outputs from other projects may also be required, or other conditions satisfied before project purposes are achieved. The use of the word output can give rise to some confusion, since it is often used to mean the physical outcome of a production process. In this context, however, project outputs refer to the tangible physical and institutional structures established through the project. It is the operation of these structures that enables the projects purposes to be achieved. Examples of project outputs include wells for the provision of safe water, factories that are established to produce a particular product, clinics providing advice and medication to expectant women, and irrigation schemes that allow farmers to increase production and employ more labour.

The project inputs are the resources required to establish and produce the project outputs. These include finance, skilled and unskilled labour, materials, capital equipment and technical assistance, among others.

The narrative summary lies at the heart of the project framework. It is particularly useful for managers in helping them to clarify project objectives and define project strategy.

Table 2.1 The project framework (typical rural water project)

	Narrative summary	Assumptions	Indicators	Means of verification
Goals	Improved health of rural population	Health education will be available regarding cleanliness, sanitation, etc Other environmental factors not contributing to ill-health	Reduction in deaths and sickness due to water-borne diseases	Hospital and clinic statistics Social Surveys
Purposes	Increased consumption of safe water	Rural population will wish to use pumps, rather than rely on traditional sources Pumps will operate satisfactorily and can be maintained	Utilization of wells Per capita consumption of water	Field surveys and observations Records of pump breakdowns and down time
Outputs	100 wells drilled, and installed with handpumps	Accessible groundwater is available in the project area Suitable locations for wells can be found.	Number of completed wells	Field inspection Project records
Inputs	Resources • skilled and unskilled labour • finance • materials and equipment	(Assumptions regarding acquisition of projects should be covered explicitly by project design and planning)	(These areas are a specific concern of project management)	

Clarity of objectives at each level is important for successful project execution: without such clarity there can be no definition of the plan and no effective exercise of management. The vertical logic of the framework also forces managers to ask the question 'in what way does this particular activity contribute to project purposes', and thus raises the possibility of other methods of achieving these purposes, in effect requiring from the manager careful analysis and definition of the project strategy itself. This may be especially relevant to adaptive, people-based projects, in which managers may have considerable latitude in determining strategy.

The causal links between inputs, outputs, purposes and goals returns us to the role of policy formulation and its link with project development touched on in Chapter 1. Policy-makers determine the developmental goals to which project purposes contribute, and policy may also determine the soundness or otherwise of the assumptions made concerning the logical causative links of the narrative summary. Project managers may well contribute to the policy debate and even be instrumental in formulating policies but, in their capacity as project managers, they are particularly concerned with the transformation of project inputs to outputs. This transformation is, in fact, the project. The project (the time-bound investment of capital to produce assets, as defined in Chapter 1) ceases when the transformation of inputs into outputs is complete. The contribution of project outputs to the achievement of purposes takes place when the assets have been created and are in operation. Project managers may be in a position to exert some influence on how this takes place but the main focus of managerial interest for project managers lies in the efficient and effective transformation of inputs into outputs.

Assumptions

The relationships between the project goals, purposes, outputs and inputs are represented in the project framework as a vertical logic from the mobilization of resources to the attainment of development targets at the sectoral, regional or national level. However, the feasibility of this logic is based on various assumptions which are also defined in the framework, in a second vertical column associated with the narrative summary. These assumptions relate first to complementary outputs from other projects which are necessary to support links in the chain of project objectives: for instance in the example, increased provision of clean water will not lead to improved rural health unless it is accompanied by a health education effort which may well be the intended output of another project. Second, the assumptions relate to conditions in the environment which must be satisfied before the causative links will operate. Thus locations for wells must be found which are suitable not only in physical terms but also socially and culturally: suitability may be affected by such factors as landownership, and traditional patterns of water collection, over which project managers have little influence, and which therefore form part of the project environment.

Assumptions are made at each level of the project framework regarding the conditions necessary for the successful transformation of project inputs into outputs, the contribution of outputs to purposes and purposes to goals. It is therefore consistent to make no assumptions at the lowest level, regarding the acquisition of project inputs, since this lies entirely within the managerial interest. This is not necessarily to say therefore that the task of resource acquisition is easy for project managers, but rather that this task is an integral part of their overall responsibility. Without inputs there can be no outputs and an essential feature of the managers' job is to acquire and control the necessary inputs.

Experience indicates that, where projects fail, the assumptions made by planners and managers have often proved invalid for some reason. Farmers fail to achieve expected yields, the volume of water available on an irrigation scheme falls below expectation, clinics remain unused for lack of trained medical staff. The main assumptions indicated in the project framework are those that establish the causative nature between the project inputs, outputs, purposes and goals. If for any reason these become invalid, there will be a profound effect on the outcome of the project.

Indicators and means of verification

The narrative summary represents the causative linkages between the main project elements in a progression that begins with resource inputs and ends with the attainment of development goals. If, however, these elements are to have more than just a theoretical basis, they need to be redefined in practical terms. The project framework attempts to do this through the identification and quantification of indicators which measure achievement at the different levels.

Thus the framework includes a column in which an attempt is made to identify the indicators that describe those elements in the narrative summary. As far as possible the indicators should be precise and quantifiable, in order that there can be no ambiguity about success or failure in achieving them. In the example the number of wells drilled and completed is a precise measure of the achievement of project outputs: the target number will of course be changing over time and the project framework must reflect this. The advantage of identifying indicators of elements of the project in the narrative summary is that it provides planners and managers with a clear set of targets at each level of the project and it ensures that progress can be measured against the targets. In this, there is a close link with the growing interest in the use of performance indicators in a wide variety of managerial situations. Indicators also make possible the comparison of project inputs with the completion of project outputs and achievement of project purposes and goals, thus providing the basis for project evaluation. They would help to show, for instance, whether a rural water supply project did actually contribute to increased health in the population. If not, causes could be identified and improvements made on the existing system and future projects.

The means of actually measuring the indicators completes the framework, again using a vertical logic. This can be used to establish the basis for a monitoring system, both for the project itself and for the subsequent operation of the system.

In some cases verification will be a straightforward process involving inventories of assets and facilities. Verifying that project purposes and goals are being achieved may, by contrast, be much harder. It will probably involve some sort of field survey and primary data collection exercise, and will also require the establishment of the baseline situation, against which improvements can be measured. The means of verification must be feasible within the context of the project: in particular, resources in terms of project inputs may often be needed if the measuring system is to be effective.

This brief review of the project framework would not be complete without mention of some of the limitations of the concept. The project framework does not substitute for the rigorous processes of project formulation and appraisal. For instance it does not describe alternatives, nor ensure that the project approach taken up in the framework is optimal. It does not describe how the transformation of project inputs to outputs and outputs to purposes will actually take place: it remains the responsibility of managers to ensure that the transformation takes place efficiently and effectively. Projects are dynamic undertakings and the project framework describes the situation at a particular time. There is a danger, as there is with many other management aids (organization charts and implementation plans, to name but two), that the framework will be drawn up once and then left untouched to become increasingly irrelevant. This danger can of course be avoided by a commitment to update the framework if the situation changes sufficiently to warrant it.

In spite of these limitations, however, the project framework remains an invaluable aid for the planner, and particularly the project manager, in encapsulating the project succinctly, identifying project approaches and strategy, assessing the relationship of the project to its environment, and in defining the key elements of the monitoring system. The description of the project framework for the rural water project has deliberately been kept simple for illustrative purposes. In practical situations the framework for even comparatively simple projects will be considerably more complicated than that shown in Table 2.1.

The project environment

In Chapter 1 the project environment was defined as the whole set of people, things and institutions which surround the project and interact with it. This concept came originally from the natural sciences, where it refers to the natural setting of living organisms at varying levels, and includes the concept of the interaction between organizations and their environment as part of their continuing survival strategy.

The biological environment of an organism relates to its physical and natural surroundings, while the environment of a project can be described in institutional, social, political and economic terms, as well as physical ones. This environment, however, is one on which the project depends for its survival in the same way that a living organism depends on its biological environment. There must be continued interaction between the project and its environment, and the project itself affects the environment just as it in turn is affected by it, so that there is a two-way relationship between them. The concept of mutual interdependence and survival demonstrates the importance of supportive linkages with the environment for the successful outcome of a project whether at the level of outputs, purposes or goals. Commentators such as Conyers and Kaul (1990) are stressing the vital role that a favourable environment plays in successful project or programme development. While effective management has a similarly vital role to play, it is undoubtedly true that a hostile environment increases the demands made on managers and diminishes the likelihood of success.

The project environment includes a whole range of factors, many of which will have a direct bearing not only on the way the project is actually implemented but also on its outputs and how it is subsequently operated. Each project will of course have its own unique environment and there is no definitive method for detailing the precise nature of the environment of a particular project. It is, however, vital that managers develop their abilities to understand the characteristics of the environment, to analyse it, and then to develop coherent management strategies for dealing with it in such a way as to enhance the chances of project success. The use of the project framework assists managers in identifying the significant elements of the project environment. Subsequent paragraphs discuss methods of analysing and coping with those elements.

Characteristics of the development environment

The idea of the project or organizational environment has been current for some considerable time in managerial thinking, both in industrialized and developing countries. It is commonly held, however, that the environment in developing countries poses special problems for managers.

Two general characteristics of the development environment apply in many situations and lead to these special problems. The first aspect is turbulence, the tendency for unpredictable and rapid change. Rapid change is a feature of the late twentieth century in societies across the development spectrum: indeed mass communication has ensured that it is very difficult to prevent changes in one area from influencing changes in another. In industrialized societies, however, the effects of such changes are relatively predictable over the short to medium period: in broad terms people know what to expect from changes in economic circumstances and advances in technology, and the reactions and adaptations of organization and society can be foreseen with some degree of confidence. This is often not the case in developing countries where sudden and unpredictable effects follow the occurrence of particular sets of circumstances. The natural environment itself can have disastrous effects on developing economies, whereas industrialized nations are, generally speaking, better equipped to forecast and handle such disasters and minimize the disruption they cause. For example, massive flooding and cyclones in Bangladesh can have a devastating impact on human well-being and economic activity, in a way which more developed economies are normally better equipped to prevent. Changes in other aspects of the environment such as economic down-turns and austerity programmes can have similar devastating impacts on political and social stability.

The second important aspect of developing country environments is that they often seem to lack resources. This is referred to in some management writings as scarcity (its opposite being munificence). It is important here not to confuse general resource scarcity with abundance in the provision of particular resources. It is certainly true that many developing countries are well-endowed with particular resources. Generally, however, they suffer scarcity of many other resources, to the extent that they also face a difficult and hostile development environment. In this environment managers are often constrained by the overwhelming need to acquire and then control the resources necessary to implement their project, be these human, financial or material, and they are compelled to devote to this task time and energy far beyond that expended by their contemporaries in industrialized societies. The characteristics of turbulence and resource scarcity are common in one form or another in many developing countries and need to be kept in mind by project managers, as a potential constraint on a successful project outcome.

A taxonomy of the project environment

When faced with the need to describe the environment of a particular project it is often helpful to categorize the factors into four broad areas:

1. physical
2. economic and financial
3. institutional and political
4. socio-cultural

It should be recognized that these areas are not in reality separate. They are part of the total system and each is moulded by the shape of the others. Nevertheless this is a helpful framework with which to start.

The physical environment

The physical environment refers to the natural setting of the project, its geology, soils, landscape, climate, water resources and ecological systems. It should also be extended to the technologies which are, or can be, utilized for the exploitation or conservation of the natural resources.

Many projects are sited in a particular location and are surrounded by a specific physical setting. Sometimes this setting is the very reason why the projects exist. A mining project, for instance, is established precisely to exploit the deposits in physical environment; agricultural projects exist to maximize the potential of the natural physical environment for agricultural production. Even projects which are not intended to utilize or transform the physical environment will have an influence on that environment and will in turn be influenced by it, in such aspects as climate, water supply, waste disposal and the like. Managers of 'physical' projects must be particularly sensitive to their physical environment, not only because it can exert an overriding influence on project progress, but also because of the increasing interest being shown in environmental protection and sustainability both in the developed and the developing world.

The physical environment cannot simply be described and assessed in terms of its natural elements. Inevitably projects are implemented in an environment in which certain levels and types of technology are currently being deployed and they will seek to use these technologies or import others. The availability and provision of these technologies is a key dimension of the project environment.

The economic and financial environment

The economic and financial environment is of obvious significance to projects. Projects utilize resources to create assets. The resources utilized have a cost and the assets created have a value. The relative costs and values, and hence the worth of the project are directly affected by the economic and financial environment within which it is being implemented. Economic and financial factors vary constantly and induce high levels of uncertainty in the process of project development. Decisions are constantly needing to be reviewed in the light of changes to the economy, budgetary constraints, foreign exchange shortages, price controls and other factors. To some extent all development activities are affected by these factors but they affect projects particularly severely, because projects are planned to be completed at a stated cost, using resources which the project must acquire for the purpose. Cost over-runs, often associated with time delays, are very frequently encountered on all types of development projects. In many instances, these are caused by the constraints of the economic and financial environment.

The institutional and political environment

Project managers need to be aware of the general institutional framework within which they are operating, the nature of the organizations with which they must interact. The general institutional framework concerns such matters as the legal systems within which they are operating and other aspects of social organization such as the land tenure and water rights system. For example, difficulties may be experienced by water projects which are implemented on the assumption that the users and beneficiaries will pay for water, in societies where there is a long tradition that water is free. It is essential that project managers have a good understanding of these aspects of the environment if they are to be able to manage their projects successfully.

In dealing with other organizations it must be borne in mind that in many respects development projects are intrusions into the established

23

organizational framework. They are often only temporary members of that framework, due to the fact that they have a defined life and, in addition, may have overlapping objectives with many of the existing organizations. There is, thus, clear scope for both mutually supportive interaction and conflict between the project and the organizations with which it must relate.

Although this relationship is a crucial determinant of project outcome it is not necessarily the formal organizations that are the key elements. Government departments, marketing and credit institutions, local councils and other formal organizations established under precise terms of reference are often of major concern, but there are a wide variety of other organizations that may be of greater significance. Studies of rural development projects have indicated that there are many less formalized and official organizations such as village level committees, political groups and religious institutions operating in rural communities; they are not as immediately recognizable as formal organizations, but they may be as important. If these organizations and their leaders are not sympathetic to the aims and activities of the project, they can be very influential in hindering and constraining it.

On a more general level politics plays a vital role in determining the progress of any project. Politics is an essential feature of human organization and permeates through all levels of society. As tools of economic development, development projects necessarily reflect the political priorities of the country within which they are being implemented. In addition aid-funded projects will reflect the political priorities of the funding agency and donor governments. There is therefore a political framework in the project environment that needs to be considered.

Project managers therefore need to be political. They will need to negotiate, lobby and use influence to gain political support for their project. The political environment surrounding the project will involve a complex web of political relationships which extends beyond individuals to organizations and geographical areas. Like the other aspects of the environment it needs to be analysed, understood and addressed, if the project is to be implemented successfully.

The socio-cultural environment

Many projects are specifically designed to develop human resources as a major objective. These people-orientated projects include most agricultural projects, health, education and welfare projects, urban housing, water supply, and rural industrialization projects. Their essential feature is that they can be implemented only through people who are not directly part of the formal project organization.

The project organization provides resources, training, services and infrastructure to the population but does not in any way control their decision-making or actions. In order for such projects to be successful, therefore, its objectives will need to be consistent with the values and practices of those people it is designed to assist. This seems obvious but it is much more difficult to bring about in practice than in theory. Assumptions about socio-cultural values and practices may not be valid for a whole variety of reasons, particularly if they are made by planners who are not closely identified with the target group. In addition, societies are complex structures which cannot easily be defined in terms of a single set of behaviour patterns and great differences are often observed within nations and inside regions within nations.

The word 'culture' suffers from a variety of interpretations. For this purpose it means the shared norms and values of the local population and the way in which it arrives at decisions and executes individual and collective action. It is vital for project personnel to be sympathetic to the local culture and to have an understanding of 'why things are done the way they are'. An illustration from a real-life example helps to illustrate this.

On the Rusitu Small-scale Dairy Settlement Project in south-eastern Zimbabwe settlers were required to produce milk from improved cross-bred cattle. A condition of the leasehold was that all heifers were to be serviced by the improved bulls kept by the project. No local bulls were allowed on the scheme to avoid the chances of heifers being serviced by them instead of the scheme bulls. Years after the project started the managers still reported that settlers were refusing to comply with the condition of not running local bulls with their dairy cattle. Ultimately this would have severe consequences for the future viability of the scheme. Following some further investigations into the matter it soon became clear why the settlers were not willing to comply.

Most of them were drawn from an ethnic group that had long been used to cattle ownership. Cattle were a major source of capital accumulation as well as milk, meat and currency. This dependency on cattle went beyond economic considerations, however, to the extent that it was a commonly held view among this group of people that the family bull that ran with the herd was the embodiment of their own ancestral history. There was an established linkage between the human ancestral line and that of the cattle herd. It was not surprising, therefore, that settlers were reluctant to abide by the condition of not running their own bulls with their dairy herds. This simple but powerful illustration indicates the magnitude of the problem facing people-orientated projects. There are no set rules and procedures for coping with these types of difficulties but it is imperative that managers are aware from the outset that people will not necessarily share their own values and that different cultures and practices may explain people's response to project initiatives.

Modelling the project environment

So far characteristics and categories of the development project environment have been suggested for use in describing and classifying its key elements. Effective project managers will, however, need to be able to move beyond understanding and analysis of the environment, to develop approaches for establishing supportive linkages with their projects. A number of environmental 'models' which define these linkages in one way or another are useful in this respect (Youker 1987).

The first is the input-output model (see Figure 2.1) which envisages the project as a process of transforming inputs into outputs, which may be goods, services, institutions, or even people (if, for example the objective of the project is to develop skilled technicians or community workers). This process of transformation is affected by regulators and competitors. Regulators in this model are those people, organizations and institutions that exert direct influence over the project and its inputs and outputs. The competitors are those that compete for project resources and indeed compete to achieve the same objectives.

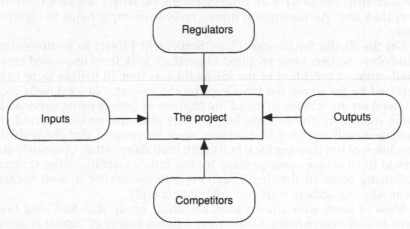

Fig. 2.1 The input-output model

The input-output model is particularly valuable in that it can be used equally effectively both for project implementation, when the facilities or systems are being created, and subsequently during the operational phase. The inputs, outputs, competitors and regulators change, of course, from one phase to the next. Thus for the implementation phase of the rural water project, the inputs and outputs are, respectively, resources and completed wells, exactly as defined in the narrative summary of the project framework. Regulators would include local and regional government, which might want to influence the siting of wells both on administrative and technological grounds, and local informal community groups. Competitors of importance might include other users of skilled and unskilled labour (agricultural harvests, or other construction projects for instance), and purchasers of construction materials.

At the operational stage the inputs become the completed wells and the output, increased consumption of clean water. While there may not be any significant regulators of this process, an important competitor would be the traditional sources of water which, though unsafe, may be more accessible to the users, and free. Identification of competitors and regulators in this way would assist the manager in establishing relationships with the competitors and regulators which are most conducive to project success.

Another model depicts the project environment in a hierarchical manner. This is a very useful way of building up a picture of the institutional framework, as it indicates the levels at which certain issues may need to be taken up. For example policy matters (such as repayment and maintenance arrangements) may need to be referred to the national level within a ministry whereas regulatory issues (such as well-siting restrictions) may be dealt with at the district level. Other institutions such as credit suppliers may have a hierarchical structure with each level of the hierachy having a ceiling on the amount of money they can lend out. Building up a hierachical model for the main institutions with which a project needs to deal can thus prove useful.

The third model is the 'actor/factor' model. This divides the project environment into

- actors – those organisations and people who can take actions which influence the project, and
- factors – which are not able to take action but simply exist as part of the environment.

Examples of actors include the organizations and institutions referred to in the previous models. In addition there may be important individual actors in the project environment such as prominent politicians or village heads whose support is vital for the smooth running of the project. Factors would include the physical elements of the environment (water table depth, nature of the rock to be drilled), the legal framework (land tenure systems), and the cultural context (patterns of household practices and social relationships). These do not react directly to the project but affect it by their existence. Distinguishing between actors and factors in the environment is useful for project managers because the two sets of elements need to be handled differently. Having first determined which are the most important in each category, managers will need to establish supportive links with the actors through an energetic process of formal and informal communication. Key factors, on the other hand, need to be continuously monitored so that managers receive the maximum amount of warning of any adverse circumstances and trends and can take any necessary or possible steps to mitigate them.

The final model is the 'degree of control' model. As indicated in Figure 2.2, this divides the environment into concentric rings based on the nature of its relationship to the project. The part of the environment directly controllable by the project is set in the innermost ring, and the project should be planned in such a way that control over these elements of the environment is specifically provided. They may be thought of as part of the transformation process from inputs to outputs and thus, strictly speaking, lie within the project and are no longer part of the environment. The next ring is that part of the environment which the project cannot control but over which it can exert a greater or lesser degree of influence. The outer ring is that part of the environment which directly affects the project but which the project itself can not influence in anyway. This is sometimes called the 'appreciated' environment.

27

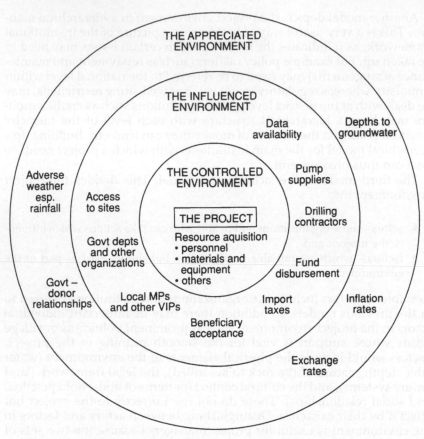

Fig. 2.2 The degree of control model

The degree of control model adds a useful dimension to managers' understanding of the environment by underlining the complexity of the relationship between it and the project. On the one hand there are situations which they control, on the other hand there are situations which they are unable to control but which greatly affect their project and in between a great range of situations in which there is some scope for action. This may be very limited in some instances and very great in others: it depends to some extent on the type of project, with resource-intensive projects requiring a greater degree of control for their success while people-oriented projects may provide more opportunities for influence than control. The degree of control model also emphasizes to managers that it is impossible for them to have complete control over all aspects of the environment which affect their project. Many commentators suggest that the real ability of managers lies not in the way they exercise control over the controllable elements of their project but the way they direct it by bargaining and negotiation through the influenced and appreciated environment.

Assessing the characteristics of the environment, describing its key elements, and then modelling in various ways their relationship to the project provides managers with a conceptual framework which enables them to develop appropriate ways of dealing with the environment. Without such a framework there is a danger that attention will be concentrated entirely on internal management processes, for which some systematic framework, however imperfect, can be developed. However, as Chapter 1 repeatedly emphazised, there is ample evidence that environmental factors are often at least as important as internal factors to project success.

References and further reading

Bryant C, L G White 1982 *Managing development in the Third World*. Boulder, Colorado, Westview.

Coleman G 1987 Logical framework approach to the monitoring and evaluation of agricultural and rural development projects. *Project Appraisal* **2** (4): 251–9.

Conyers D, M Kaul 1990 Strategic issues in development management: learning from successful experience. *Public Administration and Development* **10** (2): 127–40.

Youker R 1987 *The development project environment*. World Bank EDI, Mimeo.

CHAPTER 3

Project management and the project manager

The concept of management

Management, in the sense that it is used throughout this book, is the process of getting work done through other people by the use of human resources, materials and time to achieve objectives. Many people across a wide spectrum of human endeavour are concerned to achieve objectives using materials and time. The key distinction of management lies not merely in the attainment of specific objectives, but in attaining them through the efforts of other people. Project managers, who form the focus of this book, are concerned to achieve specific objectives, in this case project outputs, through the efforts of the other people making up the project team.

In arriving at an understanding of the constraints and opportunities of project managers in developing countries, it is valuable to understand how this concept of management has developed in industrialized societies. Prior to the nineteenth century, management in this sense was practised only by military organizations and established religion. Both these, and the expanding industries of the industrial revolution, were primarily concerned with physical resources and paid little attention to the management of the workforce as individuals, except to the extent that 'leadership' of those individuals was a prized and desirable quality. In the early twentieth century the Scientific Management movement challenged this approach. Scientific Management recognized the importance of the individual workers in the productivity of the enterprise but treated them, in effect, as part of the working machine. It also tried to improve practices by separating the planning from the execution of the work and emphasized an engineering approach to work processes. F W Taylor, the most notable exponent of Scientific Management, conducted a famous experiment at the Bethlehem Steel Company in the USA, in which he studied the working patterns of a particular manual worker in great detail. Using a mixture of changes to these working patterns (for instance, changing the design of the shovel used, and the frequency and duration of rest breaks) and financial bonuses for faster work, he was able to bring about vast increases in the productivity of this worker. Taylor's advocacy of such techniques as time-and-motion studies and piece-rate incentives stemmed from this general approach but predictably aroused the opposition of workers and the unions who resented being treated so mechanistically. Moreover, the theories began to lose credibility as it was discovered that they would not stand up to rigorous scrutiny and testing.

Scientific Management was succeeded by the Human Relations movement, which developed in the 1930s and 1940s in parallel with advances in psychology and sociology. The Human Relations movement shifted attention from the technical aspects of the workers' job to their relationship with the organization and their fellow workers. The Hawthorne studies of that period played a major role in the development of the Human Relations movement. These were originally intended to bring about improvements in productivity in a group of workers involved in intricate manual work, through a series of measures such as improved lighting designed on the principles of Scientific Management. The experiments showed, however, that the performance of the workers was related to psychological and social factors as much as to physical ones and focused attention on job and worker satisfaction, rather than the physical and mechanical components of working practices.

In turn, however, the Human Relations movement also began to lose credibility when it was found that it could not provide a coherent and comprehensive management paradigm; it has itself been overtaken by theories developed in the 1960s and 1970s, but is still influential today. These theories, which are sometimes called Systems Theories because they view work and its management as a set of interlocking parts in a complex linkage with its surroundings, were developed by workers in the field of organization and behavioural science (see Chapters 5 and 10). In parallel with work done in these areas, other specialists used new developments in mathematics and computers to develop such techniques as linear programming, risk analysis and critical path methods which go under the general heading of management science. Later in this chapter, and indeed throughout this book, the relevance and appropriateness of management theory, which has largely been developed in industrialized countries, will be examined in the context of project management in developing countries. In one respect at least, however, the systems theories, with their emphasis on the relationship and interaction between the organization and its environment, provide rich opportunities for comparative analysis. Much development effort is at present directed towards the goal of providing a better 'fit' between the project and its environment and of adopting working practices which build on the best of indigenous skills and attitudes.

The nature of project management

Chapter 1 provided a general definition of a project and drew a distinction between projects and other organizational undertakings such as the operation of systems or facilities created by projects. This distinction extends to the management of these activities. Whereas many texts deal with 'management' in an undifferentiated sense, this book discriminates between project management, its main focus, and other types of management, such as operations management and agency management. In this connection it is also worth drawing a distinction between 'management' and 'administration'. Managers differ from administrators in that they are expected to influence and direct the activities or organizations for which they are responsible. Projects are initiatives designed to make an impact on their environment and to achieve development through change: it is the responsibility of the project manager to guide the project towards achieving effective change. Administration, by contrast, implies the maintenance and smooth functioning of existing procedures, particularly when applied to the functioning of government bureaucracies.

Although project management lies within the general sphere of management, the particular nature of projects makes for some important differences from other types of management. The latter are normally concerned with continuing operations or long-term development, whereas projects, by contrast, are temporary and time-bound. This has specific implications for a number of the manager's priorities and activities, such as resource acquisition, the design of project organizations (discussed in Chapter 4) and the management of the people concerned with the project. For instance, project managers usually cannot, except in particular circumstances, develop staff for their project. They must use the human resources which are readily available to them, whereas other organizations can take a long-term view of their staff requirements and potential. On the other hand projects have a task-orientation which provides a source of intrinsic satisfaction to those working on them and also provides powerful motivation to perform well even in quite adverse circumstances. This must, however, be counterbalanced by the temporary status of the project and the feeling of insecurity that this engenders in these staff. Such insecurity may be compensated for by enhanced conditions of employment, which may be a cause of difficulty in public sector administrations.

Figure 3.1 presents a basic framework of the nature of project management, around which the remainder of the book is constructed. This has been deliberately kept simple, in order to make it as generally applicable as possible. Readers may like to extend the figure, and to add further linkages, in order to reflect more accurately the realities of their own situation. It should also be borne in mind that the relative importance of the various aspects of project management will vary from project to project. For instance, in people-centred projects based on a set of untried technologies, a major concern of project management will be a flexible project strategy, whereas in capital-intensive investments, strategy is often closely defined by project technology, so that planning and resource control will be the main priority.

The first two chapters of this book have discussed the nature of projects and described concepts which will assist project managers to define appropriate project strategies. The next part of this chapter examines the roles and functions which managers use in pursuit of project strategies. Later chapters deal with other key aspects of the project manager's work, starting with perhaps the most important, working with people within institutions and organizations. Besides the management of the project organization and the people within it, managers of projects in developing countries will also be concerned with other organizations and institutions and co-ordination with them. A particular feature of development projects in the public sector is the large number of autonomous organizations that are sometimes involved, including local government, supply services and contracting organizations as well as the main sponsoring agency.

Other concerns of project management are also shown in Figure 3.1. The resources required for project investment will generally be beyond the scope of internal generation through the organization's normal activities and will therefore need special attention in their acquisition, allocation and control. Planning and scheduling is more important for projects than for ongoing operations because existing data from previous operations are not directly appropriate to guide managers in decision-making. A further aspect of this is that it is often necessary for project managers to obtain work, goods or services from outside the organization, through the practices of procurement and contracting.

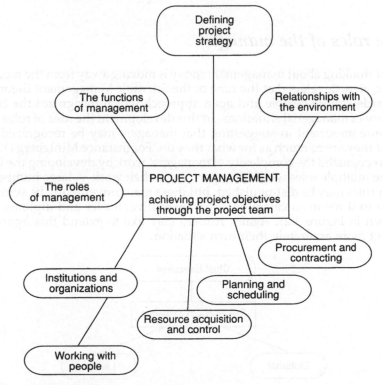

Fig. 3.1 The framework of project management

Finally, project managers also require interaction with, and support from, the project beneficiaries and a variety of different agencies or institutions which are outside their immediate control or sphere of influence. The interaction with the project 'environment' is a process from which other types of managers, for instance operational or production managers, may be to a large extent protected.

The five elements which form the major concern of project management are

1. people, organizations and institutions
2. resource acquisition and control
3. planning and scheduling
4. procurement and contracting
5. interaction with the environment.

They are likely to be present to some degree in all projects of whatever nature, though their relative importance will, of course, vary from one type of project to another. Each will be examined in detail later, but first it is necessary to consider the breadth of project managers' responsibilities and to review the various roles and functions which they may be required to perform.

The roles of the manager

Most thinking about management today is moving away from the mechanistic view developed at the time of the Scientific Management theories, towards a more flexible and open approach which recognizes the complexity of managerial situations. In this development the idea of roles has become important in suggesting that managers may be recognized for what they are as much as for what they do. For instance Mintzberg (1973) has recognized the complexity of managers' work by developing the idea of the multiple roles which they perform in their work. A large number of such roles may be distinguished, but these may conveniently be summarized in three main categories, chief executive, leader and diplomat (as shown in Figure 3.2). Again, readers may like to extend this figure to reflect more accurately their own situation.

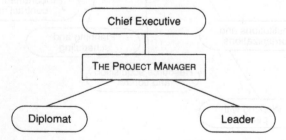

Fig. 3.2 Managerial roles

Chief executive

All projects involve the execution of a variety of activities utilizing physical and human resources to achieve specific objectives. Within a project organization someone must have ultimate authority for control of those resources and be accountable for the successful achievement of the objectives. This is usually the project manager, who plays the same role in a project promoting small farmers' poultry clubs as in one establishing a multi-million dollar irrigation scheme.

The issue of ultimate accountability sets the project manager apart from the rest of the people working on the project. Managers must accept responsibility for all the actions of project personnel who in turn reasonably expect that those actions are ultimately sanctioned by them. The chief executive role of the project manager, however, involves more than that of being accountable for the activities of the project and for providing ultimate validation for the actions of project personnel. It implies that the manager is expected to make things happen, by active intervention. It is in this role that the distinction between management and administration can most clearly be seen. The project manager cannot wait for changes to occur, but must actually create, through the project team, the changes required to achieve the project objectives. In this respect the manager's role as co-ordinator is crucial in co-ordinating the efforts of the project team and outside institutions, and in controlling and allocating resources, in order to achieve the objectives of the project.

Leader

Closely associated with the role of chief executive, but quite distinct from it, is the manager's role as leader. It is in this role that managers exert authority and influence directly over the people working either for the project or in the local environment. In their role as leaders, project managers define the ethics, norms and values of their project team, establishing the atmosphere of the organization, and the way that the various project tasks are approached. In the brief review of the history of management at the beginning of this chapter it was pointed out that qualities of leadership have been the subject of study and debate over a long period. This is still the case, though the ground for debate has shifted and modern theories of leadership pay as much attention to the followers as to the leader. Leadership and motivation are closely associated and are two of the basic skills in managing people (see Chapter 10).

Diplomat

The role of project manager as diplomat reflects the fact that a key requirement of the manager is to 'manage' the project's frontiers. Projects, by definition, are intrusions upon the existing environment, in both physical and institutional terms. The role of the manager as diplomat is to negotiate the relationship between the project and its environment.

At one level the diplomatic role is required simply to ensure adequate support for the project in terms of resources, supplies and services. At another it is necessary in order to ensure the political support, without which it is likely to fail. A common illusion of project managers is that, because the project has been planned, appraised, negotiated and agreed prior to implementation, all institutions and agencies in its environment will be supportive of the project goals and approach. This is most unlikely to be the case, because other organizations often consider new development projects as threats to their own spheres of influence. The situation is particularly pronounced in areas where many different projects, usually sponsored by different agencies, are attempting to achieve different objectives and may be competing for resources. It falls to project managers to represent their projects through contact with these other agencies.

The manager's role as diplomat requires a high level of sensitivity, good negotiating skills and a feel for the situation. A project manager who had been very successful at implementing a project at a time of great political instability and severe resource constraints remarked that the most important quality of a successful project manager was 'to be street-wise', meaning the ability to be able to understand the relationship of the project to its environment and negotiate its direction through it.

Relationship between the different roles

Clearly there is some overlap between the roles of chief executive, leader and diplomat, but it is useful to maintain the distinction in order to identify those skills and areas of knowledge which contribute to the development of each role. It may also assist potential managers to identify their roles more clearly. Managers of projects are not always aware of the significance of these roles and see their jobs as being much narrower in scope. There is ample evidence that for many development projects the problem of cost over-runs and time delays are those most frequently referred to (FAO, 1990), though these may not in fact be the most important determinants of project effectiveness. Given the visibility of cost over-runs and time delays, however, it is not surprising that managers concentrate on the chief executive role, by focusing on the technical aspects of management such as budgeting, procurement, contract supervision and physical resource control. These are immediate and time-consuming activities, without which the project cannot be implemented. The problem is that this limited role restricts managers to the implementation of project components which lend themselves to this approach such as land development, the establishment of infrastructure, service delivery systems and the installation of industrial plant and equipment. Of course the project cannot function without such tangible outputs, but an undue concentration on these activities, particularly in people-based development projects, may significantly affect their outcome. An illustration of this can be found in rural development projects, where project achievement is often initially measured through the attainment of physical infrastructural development rather than the effect the project may be having on the beneficiaries.

Completion of project outputs on time and at cost is only part of the manager's task. Central to the success of development projects is the ability to ensure that project outputs actually achieve the desired effects by creating a flexible and adaptive response to changing project situations and developing supportive links to the people and institutions in the project's environment. The manager's roles as chief executive, leader and diplomat help to achieve this.

The functions of management

Discussions of the functions or processes of management have been continuing for almost as long as management itself has been recognized as a distinctive activity. They first came to prominence in the early part of the twentieth century, at about the same time as Scientific Management was fashionable. Indeed Scientific Management and a functional approach to management have much in common. Henri Fayol (1949), for instance, described management in terms of the functions of planning, organizing, directing and controlling. Although modern thinking tends to emphasize the need to be flexible in approach and to negotiate supportive relationships with the environment as the key to success, it is useful to have in mind these basic managerial processes. Whatever the project, they will still be needed in one form or another to achieve project goals. When applying the functional approach to project management it is instructive to think of the functions as a cycle (analysing – planning – organizing – monitoring: Figure 3.3) with information from monitoring feeding back to further rounds of analysing. At the hub of this cycle is the essential managerial function of authorizing and enabling.

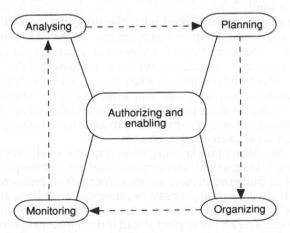

Fig. 3.3   The functions of management

Analysing

Analysing, the first stage in the management cycle, is the examination of the nature and properties of the situation in which managerial action is required. It involves definition of the activities which are needed, possible ways of carrying them out, and potential constraints. Analysing situations of a technical, logistical or administrative nature is a relatively straightforward requirement of project managers. It is more difficult to analyse the social, economic and political situations existing within the project environment, because they involve human behaviour and social organization. Project managers must at least be aware of the fact that the people inhabiting the project environment, and indeed working within the project and institutions connected with the project, may not share their own objectives, values and norms.

Failure to undertake this analysis may result in a tendency for project managers to blame individuals, groups of individuals and other institutions for the failure of projects to fulfil their objectives. Many agricultural projects, for instance, end up with the farmers being 'a problem' from the perspective of the project, because they do not repay loans, they sell fertilizer to large-scale farmers instead of putting it on the land, they 'side market' their crops and eat the improved seeds; yet all for perfectly rational reasons.

Planning

Planning is a common aspect of human activity. Even such mundane tasks as planting maize on a farm plot will require some planning. This may be just a logical thought process rather than a detailed, carefully prepared and documented plan but many human activities are based on a plan of some kind.

Project managers are involved with planning of a more complex kind, which is a fundamental part of management. Indeed it is true to say that, without a plan, there can be no management, since the plan provides the direction and framework against which management takes place. Such planning involves understanding what is required to be done to establish the project and drawing up a schedule indicating when the tasks are to be carried out. It also involves preparation of financial and resource budgets, the establishment of an appropriate organizational structure and identification of human resource requirements.

These are fundamental planning tasks required of all project managers during the early stages of project implementation, but the planning task is not completed once the project establishment or implementation phase has begun. It is a continual process, because new problems and situations arise every day which may require a new plan of action or the revision of an existing one. The ability to plan ahead is therefore an essential attribute of project managers.

Planning by itself is an important function but the ability to plan does not guarantee successful project implementation. This requires in addition the management functions of organizing, authorizing and monitoring.

Organizing

Organizing follows the process of analysis and planning and is a necessary step between planning and authorizing the plan. It involves ensuring that the physical, financial and human resources required to carry out project activities are available in sufficient quantity and in the right place at the right time. It also involves allocating responsibilities among project personnel and co-ordinating the activities of all concerned.

Organizing is a demanding task. There is need for a high level of ability to cope with the uncertainties and unforeseen circumstances that emerge, particularly during implementation. Managers can find the complex interface between the project and its environment very frustrating as their need to achieve targets through good organization is thwarted by the environmental uncertainties. It is perhaps their ability to organize yet cope with such uncertainties which will have the greatest effect in determining the outcome of the project.

Authorizing and enabling

In their role as chief executive it is project managers who are able to sanction project staff undertaking any activity. This authorizing and enabling function gives legitimacy to everyone working in the project; it also allows managers to delegate and set the bounds of responsibility for each of the individuals within it. Authorizing and enabling lies at the core of the practice of project management since it is in this function that managers 'make things happen', either directly or through the project team. While there may be some discussion as to the capacity of managers to influence situations in development projects in the public sector, successful managers are seen primarily by their ability to take the right decisions in authorizing and enabling, rather than in their skill in analysing, planning, organizing and monitoring. Indeed, while the other functions can be carried out very effectively on their behalf by members of the project team, no one can ultimately perform the functions of authorizing and enabling except managers themselves.

Monitoring

Managers need to monitor project activities to see whether or not the observed events are consistent with the plan. If not, a revised plan or other course of action must be initiated to correct the position. Through this mechanism of taking action on the basis of information collected, managers exercise supervision and control. Sometimes, as a result of the monitoring process, it becomes clear that the project is not going according to plan, or does not appear to be achieving its targets. In that situation managers need to repeat the process cycle of analysing – planning – organizing, in order to make appropriate modifications and to establish a new plan against which further management action can be taken.

Monitoring is not necessarily a straightforward process. One feature of development projects is that they are often undertaken across large areas with poor communications, so that information flows and control over project activities is not easily established. A commonly observed reaction in such circumstances is a marked reluctance by management to authorize project activities except near the centre. Agricultural projects focus on nearby farmers, irrigation schemes focus on plots close to the main facilities and health schemes focus on the main hospital rather than outreach stations. Managers must be able to monitor project activities sufficiently to ensure that they are being implemented according to plan and are achieving project objectives. Nevertheless it will be necessary to delegate reponsibility and relinquish control to some extent, particularly in widely dispersed projects, in order for changes to take place. This balance between the need to monitor and control and the need to delegate responsibility is one of the most difficult and risky areas of project management and is an aspect which will be returned to in Chapter 9.

Project management in the context of developing countries

Project managers, and indeed managers in general, in developing countries work in a different context and face a different set of problems from those in industrialized countries. This has led many observers to question the applicability to developing countries of the whole range of management concepts developed in the industrialized world, and even their general usefulness in their own context.

Fundamentally, these arguments tend to follow two main lines. One is that cultural factors play a role which does not allow for the practice of managerial skills. The other is that the environment makes it impossible for even the best indigenous project manager to operate effectively. Doubtless both factors are at work since culture and environment have a rich and varied interrelationship with one another. Certainly it would be unwise to imagine that established management practices would not need some adaptation to suit the complex situation of many developing countries.

The concept of culture is examined later in this book in the specific context of the culture of organizations. Here it is used as it is more usually understood, to mean the general framework of beliefs, norms and values which is shared in common between groups of people and which mediates on and influences their behaviour. Since managers are concerned to work with people, cultural influences are bound to be important in the practice of management. Considerable efforts have been made in an attempt to define cultural differences between developed and developing countries. Perhaps the most widely accepted framework for this is that based on work by Hofstede, which is described in Jaeger and Kanungo (1990). This framework sees five major dimensions to differentiate national cultures:

1. *power distance* the extent to which people accept the fact that power in institutions and organizations is distributed unequally
2. *uncertainty avoidance* the extent to which people feel threatened by uncertain and ambiguous situations and seek to avoid them
3. *individualism* the extent to which people feel themselves to be in a loosely knit social framework in which they are supposed to take care of themselves and their immediate families only
4. *masculinity* the extent to which the dominant values in society are 'masculine' (assertiveness, acquisition of money and things, not caring for others)
5. *abstract versus associative thinking* the extent to which people's perceptions are influenced by abstract rules and principles which they take to apply equally in all situations.

It is generally held that a 'typical' developing country will be characterized by a culture which shows high power distance, high uncertainty avoidance, low individualism, low masculinity and high associative thinking, while a 'typical' developed country will show the opposite features. Such broad generalizations will of course conceal many important differences between countries and even within countries in the same category but it is nevertheless a useful basis on which to consider how management practices may be affected by culture. For instance, managers who favour a common approach to problem-solving between themselves and their subordinates are likely to face difficulties in a society which is characterized by high power distance, where members of that society would not expect to have free and open discussions with their superiors. Project managers in developing countries will therefore need to have the skills and sensitivity to understand the basic characteristics of the culture in which they are operating, and to be able to modify their management approach and practices to fit these characteristics.

The second aspect which may require adaptation of approaches derived from industrialized countries is the environment within which development project managers are operating. In discussion of the development environment in Chapter 2 two particular characteristics were stressed, its dynamic and unpredictable nature and the lack of resources. Environments are dynamic because they change, and change very rapidly. Their unpredictability means that the logical causal relationships, on which many industrialized management techniques are founded, cannot be relied on to operate in the same way, thus negating the whole basis on which such techniques could be used. The scarcity of resources, on the other hand, places a continued set of constraints on the ability of managers to manage. For instance, the major concern of the 'street-wise' project manager described earlier in this chapter was to ensure a sufficient supply of cement for the project: all other managerial processes (analysing – planning – organizing – monitoring) took a minor role; without cement, there was no project and no amount of analysing, planning and organizing could compensate for this. This illustrates the importance of the diplomatic (negotiating) role of project managers relative to their chief executive role and technical functions.

Certainly the dynamic, unpredictable and resource-poor environment in which they work often makes demands on them not faced by their counterparts in developed countries.

Another way of looking at the importance of the environment to development project managers is through the idea of the 'locus of control'. People have an internal locus on control to the extent that they think they understand the causation of events and have control over the outcome. People with an external locus of control, by contrast, perceive that they do not understand the causation of events and have no control over the outcome. The more difficult the environment within which managers operate, the more likely they are to have an external locus of control and the less likely they are to feel in a position to 'manage' their project and the situation surrounding it.

The project's environment certainly has a very significant effect on the efficacy of particular management styles and approaches. The environment of development is generally so turbulent as to severely limit the boundaries within which managers can actually control events. Outside those boundaries, the manager is involved in processes of bargaining and negotiating, influence and persuasion which form part of a very different set of skills.

A good deal of debate now surrounds the issue of effective management in a development environment. On the one hand, one approach suggests that practices from industrialized countries have proved themselves successful and effective in their own setting and that attempts should be made to adjust the setting of developing countries to suit, particularly by reducing the unpredictability of the environment and removing resource constraints. This approach lends support to the idea that policy issues are the major set of issues facing developing countries, and that policy reforms will improve the project environment. A contrasting view is that the management theories should themselves be adjusted to take account of the setting in which they are being applied. Such a view finds support among management theories themselves, which are increasingly based on a 'contingency' view which suggests that the most appropriate course of action to adopt will take into account (be 'contingent' on) its setting.

In joining in the debate on the appropriateness of industrialized theories of management, Leonard (1987) starts by examining their basis. He describes this as

1. a commitment to collective, formal, organizational goals
2. an assumption that economics is the fundamental social process and that all other human transactions can be understood in terms of it.

Obviously these are not universally valid in any society but their applicability may be particularly limited in developing countries. It is interesting to consider where these concepts have linkages and relationships to the dimensions of national culture not just discussed. Abstract versus association thinking is obviously one area of common concern, as is individualism, and, to a lesser extent, power distance. Leonard identifies the key aspects of development management as public policy-making, leadership, general internal administration and bureaucratic hygiene. The importance of policy, the link between policies and projects, and the necessity for project managers to understand the policy framework have already been discussed. So too has the importance of leadership and the role of the manager in defining the ethics and approach of the project team. Much management theory, and particularly that labelled 'management science', is concerned mainly with the third item, general administration. This theory has developed a high degree of sophistication in dealing with problems of a technical nature, such as the planning and control of physical resources, yet these may have little relevance in the turbulent, dynamic and unpredictable environments of developing countries. By contrast, a coherent intellectual approach to the fourth item, bureaucratic hygiene, is yet to be developed. This is partly because the extent of its significance is not fully appreciated, even in developed countries. Yet all managers of projects in developing countries will acknowledge that it may be necessary on occasions to be able to circumvent rules and established procedures to make things happen. Many public administrations in developing countries try to improve bureaucratic hygiene by increasing the number of rules and procedures: perversely this may have the opposite effect since it may require more unofficial behaviour to offset its constraining effect. The extent to which corruption or other questionable practices are a necessary prerequisite of getting things done in developing countries is a matter of debate. Yet no discussion of this kind would be complete without an admission that corruption exists and that it probably affects more people in more situations than in developed countries. The existence of corruption, which by definition weakens the formal causal links on which many management approaches are based, reinforces the danger of over-reliance on these approaches by development project managers.

While the style and approaches of project development may be changing to reflect shifts in the policy framework and there are many important differences between project management in developed and developing countries, there seems little evidence that these require a fundamentally different type of management. All projects need a project manager who is accountable for project resources and who is expected to take the initiative in achieving project objectives. Such managers will play the roles of chief executive, leader and diplomat and will be expected to create change. They will also perform the functions of planning, organizing and authorizing as a means of enabling others to carry out project activities consistent with achieving project objectives. The emphasis may change according to the nature of the project, which may in turn determine to a great extent the skills demanded of the individual manager. Nevertheless, while the roles and functions of management may have been developed through an industrialized tradition of management, they are still relevant to the needs of development projects. It is certain, however, that they will need to be adapted carefully to the specific project and its environment.

The threads of this chapter are drawn together by describing the job of the indigenous manager of an agro-infrastructural project in Indonesia. This project was designed to increase agricultural production by pumping extra irrigation supplies from groundwater wells. It had thus a relatively sophisticated technological component, and was to that extent capital-intensive, but was also people-based, in that it was intended for large numbers of small farmers. The project manager had a reasonably clear set of tangible objectives (drilling and equipping a certain number of wells, and bringing supplementary irrigation to a defined area within a certain time) but was also given a degree of latitude in the approach adopted (location and size of wells, institutional arrangements for water utilization), so that the project had elements of both the blueprint and adaptive approach in its make-up. In implementing the project the manager had major concerns relating to acquiring and allocating the necessary resources, planning and scheduling the various activities, and procurement of equipment and facilities. These concerns, though by no means necessarily easy, were relatively clearly defined, and involved, to a greater or less degree, the managerial functions of analysing – planning – organizing – authorizing – monitoring. The manager was also required to operate through his own project team and to work with other institutions involved with or affected by the project (for instance, local government and the department of agriculture). While the processes of management still had some relevance to this aspect of his work, managerial roles became relatively more important as he provided leadership to his own project team, acted as diplomat and negotiator of the boundaries between his project and other organizations, and co-ordinated the variety of inputs required to achieve project objectives.

In this approach to his work, this particular manager was firmly rooted in his own culture and traditions. Indonesian society generally exhibits a high power distance which was reflected, for example, in his treatment of subordinates. While there was sometimes lengthy discussion of project matters, these were conducted in a highly formalized manner which emphasized the project hierarchy and the authority of the manager. The strongly centralized and bureaucratic nature of the Indonesian civil service also greatly affected the management of this project: interestingly the physical separation of the project from the central civil service in the capital added to the manager's authority since he was the embodiment of the civil service on the project and it was simply not possible to exercise comprehensive control of the project from the capital.

The manager's dealings with the environment took up a considerable amount of his time. In this case the environment was not particularly unpredictable but it was certainly relatively resource-poor, and much effort went into securing sufficient resources to achieve his aims. In this negotiation and bargaining was a major part of his job.

This particular manager's job has been described at some length because he is typical of many managers of development projects. Throughout this book contrasts are developed between, for instance, blueprint and adaptive projects, and between industrialized and indigenous management systems. Such contrasts are necessary to clarify thinking about these matters, but in reality many situations involve elements from a variety of different angles and approaches. In the remainder of the book ideas on these various angles and approaches are developed: effective project managers are those who know how to combine them to best deal with their particular situation.

References

FAO 1990, *Design of agricultural investment projects, lessons from experience.* Investment Centre Staff Papers Rome, Italy, FAO.

Fayol H 1949 *General and industrial management.* Pitman, London.

Jaeger A M, R N Kanungo 1990 *Management in developing countries.* London, Routledge.

Leonard D 1987 The political realities of African management. *World Development* **15** (7): 899–910.

Mintzberg H 1973 *The nature of managerial work.* New York, Harper & Row.

CHAPTER 4

Project organizations

Introduction

Projects, like other forms of collective economic and social endeavour, require people to work together in organizations. In this way they can achieve more than they can as individuals. Any discussion of project management must therefore give full treatment to organizations and the way people work within them because it is impossible to implement projects without working through organizations.

In this chapter the organization of projects in developing countries is considered, using, where applicable, insights from general management theory. Two basic themes are organizational structure (the way organizations are shaped and formed), and organization culture (their way of operating and doing things). These two themes are to an extent combined in the useful concept of organizational metaphors, which provide ways of looking at organizations as a whole by picturing them as machines, as organisms and as brains. With these ideas as a background, organizational alternatives for project managers are discussed. The chapter ends with an examination of the institutional framework of projects, and the way that project organizations must relate to other agencies and organizations which also have a stake in project development.

Organization structure

Every organization is fundamentally a structure for combining the efforts of the various organization members towards its goals; for many organizations, these goals would include, among others, the successful implementation of projects. Henry Mintzberg (1979) has identified the basic structure of organizations as the division of labour, so that each member makes a different contribution to the work, and co-ordinating mechanisms, by which the various efforts are integrated with one another.

People come together in organizations in order to pool their collective efforts, a process which requires that some undertake one kind of work and some another in an appropriate division of labour. Five basic parts of the organization can be defined in the pattern of division of labour (see Figure 4.1).

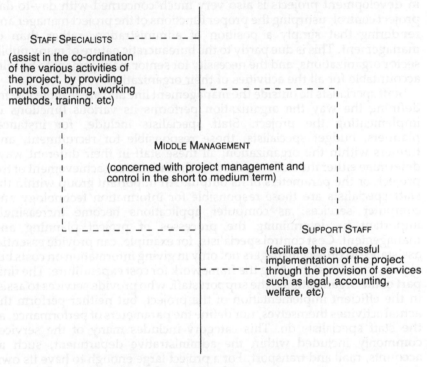

TOP MANAGEMENT

(concerned with policies, strategies and relationships with the environment eg project direction, resource acquisition, links between the project and other parts of the organization)

STAFF SPECIALISTS

(assist in the co-ordination of the various activities of the project, by providing inputs to planning, working methods, training. etc)

MIDDLE MANAGEMENT

(concerned with project management and control in the short to medium term)

SUPPORT STAFF

(facilitate the successful implementation of the project through the provision of services such as legal, accounting, welfare, etc)

FIELD AND SITE STAFF

(undertake detailed supervision and actual execution of site activities and tasks required for the project)

Fig. 4.1 The division of labour

First, there are the field and site staff who actually perform the activities required for successful implementation of the project. They include, in the case of projects, construction workers, extension staff, technologists and the like, each executing a separate part of the overall project effort. The work of the field and site staff is supervised by the line management of the organization, which can itself be divided into two identifiable levels. Middle management, such as the project manager and general supervisors, is concerned directly with the supervision and control of the work at field and site, and in processing and transmitting information both up and down the organization. Top management, by contrast, is

47

concerned more with policy direction of the organization itself, while holding overall control of the project and its strategic management. The focus of top management is on interactions with those factors, people and situations in the project's environment which have a bearing on its successful implementation. For example, top management should be concerned to ensure that overall resource availability is satisfactory for the project, and that the necessary co-ordination is obtained with other agencies with a stake in the project, while delegating detailed project control to the project manager. All too often, however, top management in development projects is also very much concerned with day-to-day project control, usurping the proper functions of the project manager and rendering that simply a position of administration, rather than of management. This is due partly to the bureaucratic nature of many public sector organizations, and the necessity for senior management to be fully accountable for all the activities of their organization.

Staff specialists lie outside the management line and are responsible for defining the way the organization performs its various functions in implementing the project. Staff specialists include, for instance, planners, budget specialists, those responsible for recruitment, and trainers within the organization; all these staff in their different ways determine either the skills required for the satisfactory achievement of the project, or the parameters of its output. An important group within the staff specialists are those responsible for information technology and computer services, as computer applications become increasingly important in determining the processes of project planning and management. Cost control specialists, for example, can provide essential assistance to project managers not only in giving information on costs but also in actually establishing the framework for cost expenditure. The final part of the organization is the support staff, who provide services to assist in the efficient implementation of the project, but neither perform the actual activities themselves, nor define the parameters of performance, as the staff specialists do. This category includes many of the services commonly included within the administrative department, such as accounts, mail and transport. For a project large enough to have its own supporting infrastructure, support staff would also include accommodation workers, catering, welfare and first aid staff, and so on.

Mintzberg's second element of organization structure is the co-ordinating mechanisms, by which the outputs of the various parts of the organization are integrated. Readers will recall that co-ordination was discussed (in Chapter 3) as part of the important role of the manager as chief executive. Mintzberg defined a number of co-ordinating mechanisms, the simplest being 'mutual adjustment', in which two or more co-workers combine their work by personal and direct contact, adjusting their own work practices as they observe those of their colleagues. Thus a health trainer may discuss and agree directly with the health specialist when, and in what form, health recommendations are required, so that the appropriate training programme can be devised. As the number of people and the complexity of work increases, co-ordination is achieved through 'direct supervision'. In this mechanism, one person becomes a 'manager', responsible for directing the subordinates in their working practices, so that their individual efforts dovetail. In the case of a health project this might be necessary if, for example, construction of health centres were also a necessary part of the project effort. It would then be the manager's job to ensure that the completion of the health centres dovetailed with the commencement of the training programmes by directly supervising the field staff responsible for each component. As organizational size and the complexity of the project work increases still further, it is no longer possible to achieve co-ordination through direct supervision. Other methods, based on the concept of 'standardization', then come into use. The most basic of these is 'standardization of work practices', in which the actual method by which the individual worker carries out the work is clearly defined and regularized. This method of co-ordination finds its fullest expression on the industrial production line and is generally not directly relevant to development projects, since a project initiative generally involves a considerable amount of non-standard work.

If such an approach is not possible, then individual efforts are co-ordinated either through standardizing the 'input skills' or the 'outputs'. In the former case the skills required to undertake the work are defined and a person with appropriate skills is employed for it. This is commonly the case in the situation where those with professional or technical training are employed. For instance, if a network schedule for a project is required as part of a project management system, a person with those analytical skills is employed to produce such a plan, which then forms the basis on which others in the team can then work in relation to financing, procurement, training, and so on.

Standardization of outputs, by contrast, defines the nature of the output required from the individual effort but does not specify how it is to be done (work processes) or what skills are needed to do it (input skills). Standardization of outputs can be appropriate either at a very simple level, for instance in defining how a security guard performs duties, or at a much more sophisticated level. It is, for instance, the method by which the overall work of a project manager is co-ordinated with other activities of the organization, the output being in this case the implementation of the project on time, to budget and of acceptable quality. Of course, the organization may also have taken into account the need to standardize on input skills by identifying the type of skills the manager would need to bring to bear on the task. The manager would also be using the co-ordinating mechanisms of mutual adjustment and direct supervision during project implementation.

Organization culture

While it is important to understand the basic structure of organizations, and to be able to interpret this structure in the particular situation of organizations concerned with implementing projects, such interpretation does not provide everything the project manager needs to know about the organization. Another important aspect is its 'culture'. The culture of an organization is the set of values and beliefs which are shared by its members, and which in turn determine its observed behavioural regularities, the feeling of the group, the dominant values and guiding philosophy espoused by it and the working norms or 'rules of the game'. Culture does not necessarily include overt behaviour patterns, which may be determined by external factors in a given situation at a particular time.

Although no two organizations are ever entirely alike, it is possible to identify a few broad categories of organizational culture which are commonly found. Handy (1983) has distinguished four such categories, of which three are particularly relevant to organizations concerned with implementing projects in developing countries (see Figure 4.2).

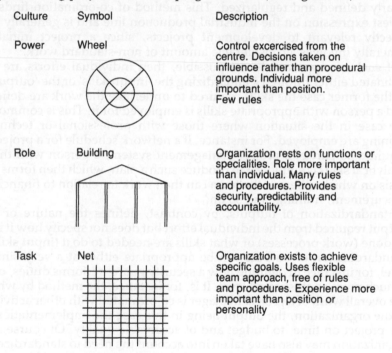

Culture	Symbol	Description
Power	Wheel	Control excercised from the centre. Decisions taken on influence rather than procedural grounds. Individual more important than position. Few rules
Role	Building	Organization rests on functions or specialities. Role more important than individual. Many rules and procedures. Provides security, predictability and accountability.
Task	Net	Organization exists to achieve specific goals. Uses flexible team approach, free of rules and procedures. Experience more important than position or personality

Fig. 4.2 Organization cultures

First, there is the 'power' culture, which is a highly political culture in which control and leadership is exercised by the individuals at the organization's centre. Decisions are taken on influence and the personal choice of these individuals, rather than procedural grounds. In a power culture there are few rules and the personality of individuals is more important than their position. Power cultures are commonly found in private commercial organizations, both in industrialised and developing countries, which are strongly associated with particular individuals, often the founders of the organization (Henry Ford is a well-known example from the motor industry). However, experience shows that power cultures are more common than might be expected in public organizations in developing countries (Franks, 1989), because organization members are often tied to their leaders through an intricate system of links and relationships which results in a measure of personal rather than procedural imperatives in decision-taking. Moreover, large organizations in bureaucratic public services often become very unwieldy and cumbersome, so that progress is achieved only when influential individuals take action which may often be outside their organizational competence. These actions, if continuously repeated, will develop a power culture round that particular individual.

Second, in contrast to the power culture is the 'role' or bureaucratic culture. Characteristics of bureaucracy are

1. specialization of jobs, which become continuing posts for which suitably qualified individuals are recruited
2. a hierarchy of authority, composed of clearly defined levels
3. impersonality, which is seen both in the administration of rules, and in selection and promotion.

In a role culture therefore, the organization rests on functions or specialities defined by the organization's activities (operations, input supplies, research, administration, etc). The organization works through an established system of rules and procedures, in which the post or role is more important than the personality of the person holding the post. An organization with such a culture provides predictability and accountability to its consumers, and security to those who work in it.

Bureaucracy has been the subject of a great deal of study and debate since its classic analysis by Weber (1947), not least in developing countries where it is often seen as an important constraint to development. Although it is often a pejorative term when applied to organizations, it is important to stress that there are many situations when the organization's customers value precisely the characteristics of predictability, impersonality and accountability which it provides. On the other hand, a bureaucratic organization is often unsuitable as a vehicle for implementing projects because it is inflexible, and because it is concerned to discharge its duties and functions correctly, rather than achieve goals and objectives.

The third type of organization culture which may be distinguished is the 'task' culture. The task culture exists to achieve a specific goal, or set of goals, of which the successful implementation of a project is an example. It is comparatively free of established rules and procedures, but instead takes whatever action is thought to be necessary to achieve the goals. In a task culture, expertise is the key factor in determining an individual's fitness for a job, rather than personality (as in the power culture) or post (as in the role culture). A task-orientated organization therefore tends to work in flexible teams, which change as the demands of the work change. As a task culture is particularly appropriate to project work, it is commonly found, for instance, in consultancy organizations which operate mainly in that way.

51

The three cultures of power, role and task go a considerable way to explaining many types of organizations and are useful in helping project managers to understand why some organizations may be successful in implementing projects, while others are not. In the situation of developing countries it would be unwise to think that these are sufficient to understand fully the working of project organizations. In particular, a feature of developing countries is the extent and importance of personal links, not just between family members, but spreading out to cover much larger groups. This concept has already been touched on briefly in Chapter 3 in the discussion of the dimensions which differentiate national culture. One of these dimensions is individualism, which tries to describe the extent to which people feel part of a large group. Broadly speaking, industrialized societies have, at least until recently, put more stress on individualism than developing societies, which have often put a high value on membership of large groups, with its ensuing rights and obligations. While the power culture, as so far developed by Handy and others, has elements of personal linkages within it, these are based on exploitative relationships (the boss at the centre uses a particular individual for a particular job because that appears advantageous: the individual complies in the hope of reward and personal advantage). Personal links of the type found commonly in developing countries seem, by contrast, to be more often supportive than exploitative.

With this reservation against oversimplification, what factors are there which would affect the culture of a particular project organization? Organizations are complex entities and there are many such factors but several important ones can be picked out. Perhaps the most obvious point to make is that, as organizations increase in size, it becomes increasingly difficult to operate them on the basis of personal choice and decision (the power culture) or goal-orientation and fitness for the job (task culture). Instead, it is necessary to turn to rules and procedures and an element of impersonality, thus developing, to some extent at least, a bureaucratic approach. Similarly, ownership is an important parameter determining culture: organizations in the public sector are likely to be role-orientated because bureaucratic procedures are better suited to the demands for impersonality, predictability and accountability by the consumers. Private organizations, by contrast, may well be identified with a strong and powerful individual, particularly in their early formative years. The organization's goals and functions will also be a significant factor. If it exists to provide a continuing, uniform service such as output production and maintenance in a stable environment, then a role culture will be appropriate, whereas an organization pursuing diverse and changing goals (including projects) will adopt a task culture. The environment itself is also a significant factor: whereas role cultures are appropriate in stable environments, they are insufficiently flexible and responsive to the dynamic and turbulent environments of many developing countries. Adaptability in such a situation may be achieved through a task culture but the power culture, with its emphasis on fast and firm decision-making may be even more appropriate.

People, too, differ in their attitudes, behaviour and feelings and some are happier working in a routine and regulated environment (role culture) while others thrive on change and unpredictability (task or power culture). Indeed national culture and characteristics may play an important part in determining a person's preferred organization culture; for instance, people from a strongly authoritarian society (high 'power-distance', in the terminology of Chapter 3) may feel uncomfortable operating in a task culture, where expertise rather than power or position confers status.

Finally, in thinking about organization cultures, it is well to remember that organizations are not monolithic or uniform. Different parts of the organization may have different cultures, those concerned with steady state activities such as maintenance having a role culture, those concerned with specific goals, such as the implementation of projects having a task culture, and those concerned with the establishment of policy and relations with the environment having a power culture. Moreover, cultures will change with time as the organizational functions change. In the context of project development this would be most notable at the change-over from project phase (where a task culture is appropriate) to operations phase (where a role culture is more appropriate).

Images of organization

A useful way of synthesizing the concepts of organization structure and organization culture is through organization imagery or metaphors (Morgan 1986). Three of these images, the organization as a machine, the organization as an organism and the organization as a brain, are particularly relevant to the consideration of project organizations in developing countries.

The image of an organization as machine suggests the idea of a perfectly formed and functioning mechanism, processing inputs into output. Each part of the mechanism is exactly fit for its role and operates in an impersonal and predictable manner. In this image the people who fill the posts of the organization are no more than parts of the machine, each discharging their duties as prescribed, without reference to personal wants or needs. Such a mechanistic method of operating is, of course, bureaucratic by its nature, and was for instance an integral part of Weber's understanding:

> Precision, speed, unambiguity, knowledge of files, continuity, discretion, unity, strict subordination, reduction of friction and of material and personal costs – these are raised to the optimum point in the bureaucratic administration. (Weber 1947)

The words themselves contribute to the overall image of the organization as a mechanism.

It is useful for all those concerned with development to speculate a little on the future of the bureaucratic form of organization in public sector administrations. Such forms and methods of operation are very widespread in many developing countries, partly as a legacy from preceding colonial administrations, and are seen by many commentators as major constraints to successful development. There is therefore a feeling that it is important to develop organizations which retain the positive features of bureaucracy (efficiency, fairness, predictability) while at the same time being more open and responsive to the needs of its customers. The National Irrigation Administration of the Philippines is a case in point. Korten and Siy (1989) describes how this has evolved from a centralized, mechanistic, bureaucratic organization to one which is open and flexible in meeting the needs of its clients. This evolution has come about partly through changes in the institutional framework within which it operates (particularly in relation to its financial arrangements) and partly through a positive effort to change its organizational culture.

The second, and contrasting, image of organization is that of the organism. In this image the concept of the environment is of major importance: the organization is viewed as a living being which coexists with its environment, mutually exchanging all the necessary inputs and resources for survival and growth. In the case of project organizations, this exchange is concerned primarily with information and resources. If the organization cannot get the right information and the necessary resources from the environment, then the project will be unsuccessful and, in turn, the organization itself may be under threat. If, on the other hand, the organization exchanges information and resources successfully with its environment, then the project will be successful and the organization will prosper and grow as the services that it provides are valued by its customers and clients. Such symbiosis between the organization and its environment is taken up in 'contingency theory', which suggests that the most effective organizational structure and management style take account of the particular conditions pertaining, both within the organization (resources, skills, goals, etc) and outside it (opportunities, threats in the environment). A consideration of the organization as organism also brings us back to the idea of an open system in dynamic contact with its surroundings, in contrast to the idea of organization as machine (bureaucracy) where the boundaries between the organization and its surroundings are carefully controlled and stabilised, as far as possible.

Finally, moving beyond the idea of organizations as organisms is the image of them as information-processing brains, which not only exchange information and resources with the environment but can also learn from experience and use that learning to alter their behaviour in order to grow more successfully in the future. Once again there are links with earlier images, through the concept of the use of feed-back to guide future action, as organisms learn their most successful behaviour as a result of the reactions of other organisms in their environment. In the case of organizations concerned with implementing projects, however, the concept of the organization as a brain means more than simply reacting to the environment but also guiding and adapting it, processes in which the project manager plays a key role.

Organizational alternatives for projects

The concepts of organization structure and culture provide a framework within which to look at the practical alternatives for organizations concerned with implementing projects. It is possible to distinguish here two alternatives: the adaptation of an existing organization to implement the project in addition to its other activities, or the establishment of an organization specifically to implement the project. A third option – sometimes called the matrix organization – takes on some of the characteristics of both alternatives. On the one hand, therefore, project owners may choose to implement projects using an existing organization. This is very likely to be structured on functional lines, in which staff are grouped according to specialized functions (planning, administration, etc). A functional organization naturally takes on many of the characteristics of a bureaucracy, including specialization of jobs, clearly defined hierarchical patterns of authority and responsibility, and a system of rules for the performance of the work. Such a structure makes efficient use of special skills and expertise by placing all those who hold them in a similar part of the organization, so that their individual efforts can be deployed to maximum effect. By grouping together, the individuals provide one another with technical and psychological support, preventing them becoming out-of-touch or isolated. Moreover, a functional organization provides an obvious career progression, in which the individuals can see themselves moving upwards as they gain experience and skills. Organizations grouped on functional lines can be, and often are, used for managing the implementation of projects. In such situations, however, the implementation may suffer because those who work on the project by providing inputs to it may not be sufficiently project- or task-orientated but will rather give priority to the demands of the department for which they work. Moreover, it is not clear who has the ultimate responsibility for successful project execution, as all staff members also have a responsibility to their own functional departments. In addition, projects require flexibility and adaptability, in order to cater for new and unexpected problems, whereas functional organizations tend to be inflexible and hierarchical in decision-making.

Figure 4.3 shows the simplified organization structure for a public works department in a Pacific island economy. Such a structure is typical of many in the public service and shows the division of the organization into its two major functions:

architecture/building   engineering/roads

with the chief architect and chief engineer at the same level, each answering to the director.

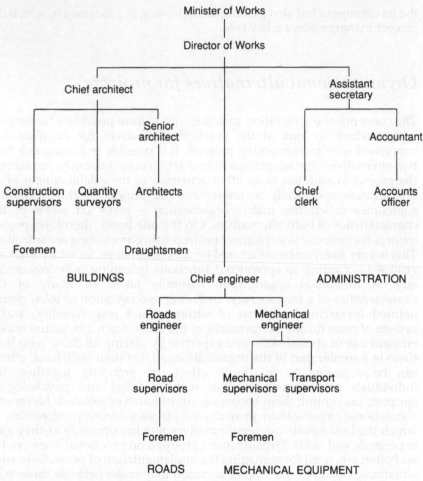

Fig. 4.3 A Functional organization

A public works department structured in this way can be used to implement appropriate projects such as the construction of an access road and a market in a rural area but each part of the organization will tend to concentrate on the areas for which it is responsible, without necessarily taking into account the contribution of that part to the overall project. Moreover communication and integration at the operational level between the engineers and architects concerned with design and construction supervision will be difficult since, as Figure 4.3 shows, their normal lines of communication are vertical rather than horizontal. Thus effective project development may be jeopardized by the traditional functional activities of the various departments.

To avoid such problems, the project owner can decide to set up a separate, largely self-contained organization specifically for the purpose of implementing the project (sometimes called a 'projectized' organization). Such a structure brings together in one location all the human resources required to implement the project. For larger projects, the projectized organization will itself be structured on functional lines, with, for instance, all those concerned with planning working together in one organizational location.

The structure of a typical project organization is shown in Figure 4.4, with the project manager supervising sections concerned with planning, land development, infrastructure and administration. Other project managers concerned with implementing other projects within the organization have similar sections, which will in turn be supported by functional departments within the head office concerned with the various specialities.

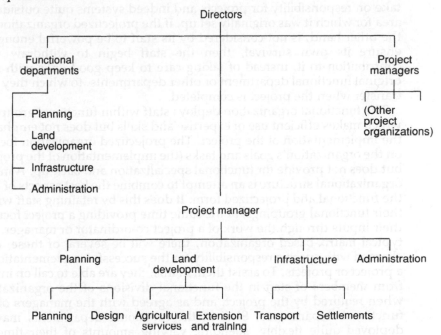

Fig. 4.4 A project organization

The projectized organization is different from the purely functional organization because all who work in it have the implementation of the project as their primary responsibility and main aim. It thus provides a task-orientation and a unity of command which should in general make for efficient implementation. However, a projectized organization has its own drawbacks. First, by withdrawing specialized resources from their functional location, the advantages of centralized grouping and allocation of these resources are lost, leading perhaps to a need for duplication or a failure to provide the necessary back-up and support. For instance, if a design specialist is assigned full time to a particular project organization, that person cannot make contributions to the other work of the design department, nor can he/she receive immediate support and back-up in

57

his/her work. Second, projects are, by their nature, temporary undertakings which eventually come to completion. At that time, the rationale for the organization's existence ceases and it should be disbanded. This, however, causes a great deal of insecurity to the staff in the organization, an insecurity which is compounded by the feeling that there is no clear career path for them to follow. If the projectized organization has become sufficiently powerful, it then looks for other activities to undertake. For instance, the Mahaweli Authority in Sri Lanka was originally set up to implement the projects related to the Mahaweli irrigation and power programme. As these came toward their conclusion, the authority, which had become very powerful through a combination of political influence and control of massive investment resources, began to take on responsibility for projects and indeed systems quite outside the area for which it was originally set up. If the projectized organization, on the other hand, is not considered by its staff to be powerful enough to ensure its own survival, then the staff begin to withdraw their contribution to it, instead of taking care to keep good links with their original functional department or other departments, to which they may transfer when the project is completed.

The functional organization deploys staff within functional groupings which makes efficient use of expertise and skills but does not emphasize the implementation of the project. The projectized organization focuses on the organization's goals and tasks (the implementation of the project), but does not provide for functional specialization and back-up. A matrix organizational structure is an attempt to combine the best aspects of both the functional and projectized form. It does this by retaining staff within their functional groupings, at the same time providing a project focus to their inputs through the work of a project co-ordinator or manager. In a typical matrix-based organization, there will be several of these, each charged with specific responsibility for the successful implementation of a project or projects. To assist them in this, they are able to call on inputs from members of staff in the functional divisions of the organization, when required by the project, and as agreed with the managers of the functional departments. Staff in the functional departments may be deployed quite flexibly, spending varying amounts of their time on different projects and making inputs at different levels as required. The matrix structure superimposes a horizontal project responsibility on a vertical functional responsibility, thereby creating the network appropriate to a task culture. In this type of organization the requirements of the project determine the inputs and resources required from each member of the project team, but this may be done on a flexible, changing basis, as the project progresses and the requirements accordingly alter.

A matrix structure for a project in the process industry is shown in Figure 4.5. The project manager co-ordinates inputs to the project from three functional areas (the Works, Chemical Engineering, and the Process Technology Departments), the inputs being provided varying both in quantity and speciality as the project progresses. If necessary, the project

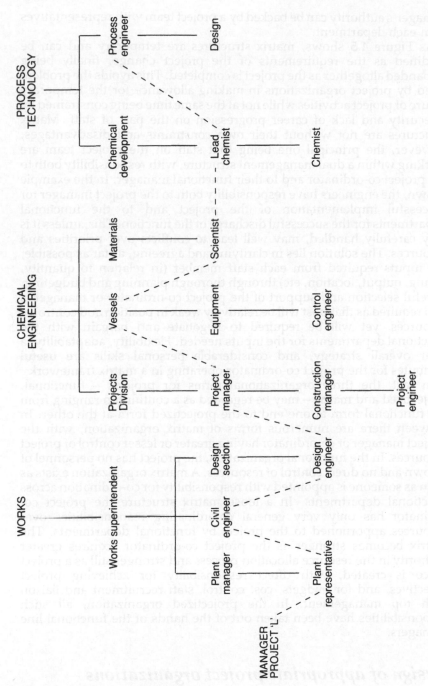

Fig. 4.5 A matrix organization

manager's authority can be backed by a project team with representatives from each department.

As Figure 4.5 shows, matrix structures are temporary and can be modified as the requirements of the project change, finally being disbanded altogether as the project is completed. This avoids the problem faced by project organizations in making allowance for the temporary nature of project activities while not at the same time being constrained by insecurity and lack of career progression on the part of staff. Matrix structures are not without their own constraints and disadvantages, however, the principal one being that staff on the project team are working within a dual management structure, with responsibility both to the project co-ordinator and to their functional manager. In the example shown, the engineers have responsibility both to the project manager for successful implementation of the project and to the functional departments for the successful discharge of the function. This, unless it is very carefully handled, may well lead to conflicts over priorities and resources. The solution lies in clarifying and agreeing, as far as possible, the inputs required from each staff member (in relation to quantity, timing, output, location, etc) through thorough planning and budgeting. Careful selection and support of the project co-ordinator or manager is also required as that post will be relatively weak in position and control of resources, yet will be required to negotiate and bargain with the functional departments for the inputs needed. Flexibility, adaptability, a clear overall strategy, and considerable personal skills are useful attributes for the project co-ordinator operating in a matrix framework.

In fact, the three organizational forms for projects – functional, projectized and matrix – may be regarded as a continuum ranging from the functional form at one end to the projectized form at the other. In between there are numerous forms of matrix organization, with the project manager or co-ordinator having greater or lesser control of project resources. In the functional organization, the project has no personnel of its own and no direct control of resources. A matrix organization exists as soon as someone is appointed with responsibility for co-ordination across functional departments. In a loose matrix structure the project co-ordinator has only very general co-ordinating responsibilities over resources apportioned to the project by functional departments. The matrix becomes stronger as the project co-ordinator acquires greater authority in the resource allocation process and stronger still as a project office is created, with direct responsibility for achieving project objectives, and for budgets, cost control, staff recruitment and liaison with top management. In the projectized organization, all such responsibilities have been taken out of the hands of the functional line managers.

Design of appropriate project organizations

Discussion of project organizations in this chapter so far has focused on structure and culture because these are fundamental to an understanding of the nature of organizations. In practical terms, however, design of an organization for managing the implementation of a project depends on three linked and inter-related processes:

1. identification of the activities and functions required
2. selection of an appropriate organization structure
3. design of job positions.

The activities and functions which it is intended to carry out are the natural starting-point for determining the actual shape and characteristics of a suitable organization. In the case of projects, identification of the activities and main organizational functions follows naturally from the analysis of the project's objectives, its various components and methods of achieving them. Such a process is carried out in great detail for implementation planning based on critical path analysis, when it is known as the work breakdown structure (WBS), to be discussed in Chapter 6. For organization design it is necessary only to identify the main sets of activities covering both physical tasks such as construction, and institutional processes such as establishment of credit supplies, provision of training and the like. These major sets of activities when grouped together appropriately and associated with the necessary support activities (including administration and financing) become the main functions of the project organization. The same sets of activities (or functions) are likely to recur in some form or other in many projects (eg design, procurement, construction, relevant institutional processes, administration, finance).

Selection of an appropriate organization structure requires a choice between the alternatives of a functional, projectized or matrix type. In making this choice, factors of importance will include, among others, the size and duration of the project and available expertise within the organization. The bigger and longer the project, the more likely it is that a separate project organization will be set up to handle it, the more so if staff resources are readily available. If, on the other hand, the project is small or of short duration, then it is likely that existing organizational structures will be retained and the project will be implemented through the functional organization, or with a project co-ordinator operating a matrix system. The matrix structure will be useful in the situation where skilled resources are constrained and it is necessary to allocate them among several projects. They are also particularly appropriate for projects aimed at development of human resources and institutional strengthening, rather than the creation of physical assets, because they aim to work within existing systems and organization structures.

The project activities and organization functions will determine broadly the job positions in the organization, but these will also be affected by organizational structure and decisions on such matters as the range of tasks to be undertaken by an individual position of department. (For a discussion of the derivation and definition of project activities see Chapter 6.) In organizational theory this is known as the degree of specialization. For instance in an agricultural project is the scope of work such that it is necessary to have different posts for agronomic trials and extension or can these be filled by the same person? Similarly, for a construction project can procurement be included within construction, or does it require a separate post with specialized skills?

The next important step is to assess the number of staff required within the organization. This should be done initially on the basis of first principles, by looking at the tasks and duties required of each position and the time required to perform them. At the most basic level, for example, reasonable estimates can be made of how long it takes a surveyor to survey a line, or how many health workers can be trained in one batch and how many trainers are required for this. More commonly, of course, the duties of particular positions cannot be so clearly defined and skill and judgement will be required to determine suitable staffing levels. In assessing total staff requirements it is of course necessary to make a trade-off with project duration. Within reason, the more people employed on the project, the shorter the time required to implement it, but it is necessary to guard against having too large a staff which then operates below maximum effectiveness because of organizational bottlenecks and resource constraints. Availability of staff is likely to be a major constraint in effective organization design; often public sector organizations attempting to implement projects in developing countries face a chronic shortage of staff. Even when staff are available, levels of training may be inadequate, and many projects are too short in duration to allow for adequate staff training. In projects, as opposed to continuing operations there will not usually be opportunity for staff to undergo training, unless special provision (both time and money) is set aside for it. If training is a major element of a project, then its implementation will take proportionately longer than if previously trained staff are deployed.

Supervision requirements will need to be determined. These will relate to the existing authority system of the public service and the organization but they will also be affected by the optimum number of people directly responsible to a single position. The modern tendency is to favour a 'flat organization', thus reducing the hierarchy of supervision. Supervision, however, provides one of the means by which effective co-ordination between the various parts of the organization is obtained. If the degree of supervision is limited, then it must be substituted by other effective co-ordinating mechanisms. In the situation of project development with its necessity to find solutions to new problems, the mechanisms of standardization are likely to be of only limited value, thus putting proportionately more emphasis on mutual dialogue and adjustment. This in turn requires an effective communication system and a task-cultured organization with its emphasis on flexible and supportive working relationships rather than the hierarchical and inflexible framework of the typical public sector bureauracy. Project managers also need to give thought to how they will provide the necessary degree of supervision without stifling initiative, enterprise and mutual collaboration on the part of their subordinates, which is such a necessary element in successful project implementation.

Organizational charting

Once the project organization has been designed, it is useful to represent its structure and working relationships in a generally available manner through some form of organizational charting. Conventionally an organization diagram, or 'organogram', is used to describe the structure of an organization. Carefully produced organograms may be found on the walls of any large organization: they are the first things to be handed to a visitor who asks about the structure and workings of an organization. Organograms depict the positions and groupings of the people within the organization in traditional hierarchical terms. They also show the system of formal authority which underpins the organization, answering the question 'who answers to whom?' They are necessary to understand the workings of an organization, but are not sufficient by themselves, especially where projects are concerned, because

1. they do not show lines of informal communication
2. they do not show the extent of responsibilities
   of each position with respect to individual activities
3. they cannot easily depict relationships between individual posts in different organizations
4. they do not illustrate the dynamic and changing relationships in organizations as projects are implemented.

The last point is particularly important for project managers in the public sector, who are likely to be working with many different organizations, including other public sector agencies, consultants and contractors.

Additional information of this type can be provided by a 'linear responsibility chart' or its simpler form, the 'dot diagram'. These charts consist of a matrix. The column heads are the important positions or roles of the organization. The row heads are the separate activities which must be undertaken to achieve the objectives of the organization. The intersection of each row and column shows the responsibility of the particular organization position (column) for that activity (row). In the dot diagram, a dot indicates the involvement of that position in that activity, but without specifying different degrees of responsibility. In the linear responsibility chart, a series of different symbols or letters at the intersection points indicate differing responsibilities, for instance

E executive responsibility
S supervising responsibility
C to be consulted (before action)
N to be notified (after action)

An example of a linear responsibility chart is shown in Figure 4.6. Such charts are particularly useful in providing visual information on links between different organizations concerned with the same project and on whether

1. delegation is adequate to avoid decision bottlenecks
2. the level of supervision is effective and appropriate to each activity
3. communication is adequate but not excessive
4. work loads are well distributed and also not excessive

Linear responsibility charts reflect the dynamic and flexible nature of projects more effectively than a conventional organogram and can show a range of responsibilities toward a particular activity. Like other project management tools, however, they may become out-of-date as the project progresses and they will themselves need revision from time to time.

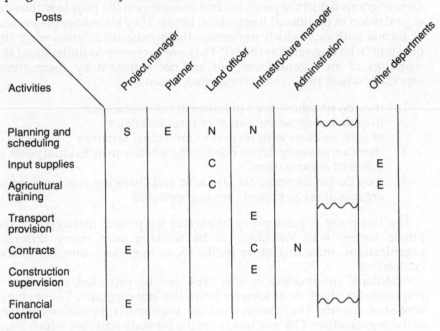

Activities	Project manager	Planner	Land officer	Infrastructure manager	Administration	Other departments
Planning and scheduling	S	E	N	N	〰	
Input supplies			C			E
Agricultural training			C			E
Transport provision				E	〰	
Contracts	E			C	N	
Construction supervision				E		
Financial control	E				〰	

Key E executive responsibility

S supervising responsibility

C to be consulted (before action)

N to be notified (after action)

Fig. 4.6 Linear Responsibility Chart

The project manager and the institutional framework

The position and role of the project manager will of course need special consideration in the design of a suitable organization for project implementation. In this consideration two factors are of particular concern: the relationship of the manager to the project owner, and the wider institutional framework within which the project is being implemented. First, managers in the public sector are normally acting on behalf of the project owner or client and are thus required to use a different set of skills from those of the manager of a private client or contracting organization. Usually the organizations of the two parties will reflect the fact that owners, and their managers and organizations, are mainly concerned with financing, project quality, overall cost and particularly relationships with the environment, while private organizations and contractors are concerned with planning, resource allocation and control. Public sector project managers will still be faced with decisions on organization design regarding the most appropriate functions, a suitable organizational structure and definition of job positions, and they will still seek to generate a task-orientated culture within their organization, but the parameters of their organization design are likely to be different from those for a private sector manager or contractor. Part of the discussion of the appropriateness of management techniques from the industrialized countries to the problems of project management in developing countries centres on problems of organizational effectiveness. Blunt (1990), in an interesting reflection on contingency theory, suggests that there is convergence across both developed and developing countries in what he defines as the structural imperatives of organizations. These include a clear definition of mission, goals and strategies and a functional alignment of the organization within these, and the need for co-ordination and control, accountability and role relationships, planning and communication, performance appraisal and commensurate rewards, together with effective leadership from the manager. Blunt believes that these are not, on the whole, contingent on the particular situation, but apply in different degrees in most settings in both developed and developing countries.

While the organizations which are established for project implementation are vital to its success, it is also necessary to consider the institutional framework within which these organizations operate. Much is made of the necessity for institutional improvement in developing countries, without a clear understanding of what this involves. In this chapter organizational structures and cultures have been considered. These, however, are only part of the institutional framework which includes, in addition, the relationships between organizations and the people who work in them and the procedures which mediate those relationships. Thus project managers may need to be involved with such matters as landownership systems, social structures, tax systems and the like. It must also be recognized that a major feature of managing projects for the public sector in developing countries is the requirement to work with different parts of the same organization and ministry and very often with completely different ministries and authorities, with whom the manager has no direct organizational link. Integrated Agricultural Development Programmes, involving investment in rural infrastructure such as roads, the provision of credit through the financial system, together with improved agricultural inputs through the supply services, are a case in point. These, too, require some form of matrix management with co-ordination across functions, in this case the individual ministries or departments with ultimate responsibility for the various functions. The normal mechanism by which such co-ordination is facilitated is through the steering or project committee, which is established to provide a mechanism for joint communication and decision-making by senior members from each agency concerned. Often the project manager is made directly responsible to the steering committee, though in other cases it has only an advisory function.

Many writers have made reference to the fact that co-ordination is an overworked concept which can often provide a rationale for inaction. Honadle and Cooper (1989) have taken the concept in a more positive direction and make some suggestions for more effective co-ordination, first by a distinction of the institutional framework into areas where the project manager has control, areas where the project manager has influence, and areas where the project manager can only 'appreciate' – in other words, seek to understand – the institutional relationships, without being able to influence them in any way. Honadle and Cooper then go on to identify possible actions in respect of co-ordination, ranging from information-sharing (the most passive) through resource-sharing to joint action (the most active). Such co-ordination comes about partly through the establishment of steering committees, among others, which allows these various mechanisms to operate.

Honadle's and Cooper's concept of the institutional framework as being controlled, influenced or appreciated borrows, of course, from the various approaches to analysing the project environment which were discussed in Chapter 2. Indeed the institutional framework is a very significant part of the project environment, in many ways perhaps the most significant since it is the least susceptible to technical analysis and treatment. Successful project managers are often those who are most astute in their political dealings with the other agencies with which their project interacts, through the institutional framework.

References and further reading

Blunt P 1990 Strategies for enhancing organisational effectiveness in the Third World. *Public Administration and Development* **10** (2): 299–313.
Franks T R 1989, Bureaucracy, organisation culture and development. *Public Administration and Development* **9** (4): 357–68.
Handy C B 1983 *Understanding organisations*. Harmondsworth, Penguin.
Honadle G, L Cooper 1989 Beyond co-ordination and control: an interorganizational approach to structural adjustment, service delivery and natural resource management. *World Development* **17**: 1531–41.
Korten F, R Siy 1989 *Transforming a bureaucracy: the experience of the Philippine National Irrigation Administration*. West Hartford, Connecticut, Kumarian Press.

Mintzberg H 1979 *The structuring of organisations*. Englewood Cliffs, New York, Prentice-Hall.

Morgan G 1986 *Images of organisations*. London, Sage.

Weber M 1947 *The theory of social and economic organisation*. London, Pitman.

CHAPTER 5

Skills of management

Farhad Analoui

Introduction

It is generally recognized that an organization's development, whether a project or otherwise, is dependent on the effectiveness of management or those with whom the strategic responsibility and decision-making rest. Managers are expected to possess skills, knowledge and competences which facilitate smooth and efficient operations. Also, they are required to manage people and tasks in order to ensure the success and survival of their organizations in the face of an increasingly complex technological, socio-economic, cultural and economic environment.

This concern for effective management is shared by practitioners, as well as management educators and trainers. An understanding of what managers do and what range of managerial skills and knowledge they require has been the result of a parallel development, within diverse interdisciplinary fields. The disciplines involved range from organizational studies, sociology, industrial psychology, personnel and marketing to economics and accountancy. The development of management skills can be traced over earlier decades and analysed to arrive at a current understanding of the art of management.

This chapter briefly considers five major categories of skills – functional, task-related, information processing, role-related and people-related – in order to demonstrate to the reader how and through what processes present-day skills have been developed, expanding on the issues raised in Chapter 3. In doing so, the underlying assumptions and reasons which have led to the perception that, for example, managers ought to be 'rational', or that they need to possess certain qualities in order to be effective, will be examined. Thus, in the form of a taxonomy the prescribed abilities (as well as perceived technical', 'functional task',`people', `information processing' and other role-related' managerial skills), knowledge and values will be outlined and each dealt with in some detail.

Much of the following discussion is based on the findings of research in the field of management development within developing countries, which results in a model incorporating a balance of the essential managerial skills and knowledge necessary for the effectiveness of development project managers.

Development of managers and required skills

Since the early days of industrialization the question of what makes an organization more effective has been the main concern for practitioners in industry, management theorists, educators and trainers as a whole. The desire for increased productivity and the prompt achievement of the organization's goals and objectives gradually and inevitably directed the attention of the theorists and educators to a special group of people in work organizations – the managers. Managers were identified as the agents with the overall responsibility for the realization of the organization's objectives. Therefore, management education and the development of their potential, abilities and skills which were thought to be necessary for getting the job done became the focus of attention.

Rational manager

With the emergence of the Weberian concept of 'bureaucratic organization', work organizations which were typically characterized by their pyramidical structures with management placed at the top of their hierarchies began to gain much importance. The hierarchy of the 'bureaux', offices which managers occupied and the prescribed roles that each had to play in relation to the achievement of the organization's objectives, became a major preoccupation for theorists, authors and practitioners alike.

Managers were viewed as those whose special skills and knowledge, particularly their 'rationality', were assumed to be the necessary ingredients for maintaining control and authority over the processes of work, people and their work relationships. Managers, by the very nature of their position within the organization structure and the fact that they occupied 'bureaux' which vested them with authority and power, were burdened with the dual responsibility of managing people and work.

From this perspective, the nature of the managers' job and responsibilities was clearly prescribed and, therefore, it was possible for it to be compared with others within the same organization.

In the early days of the development of management, it was generally believed that what managers needed in order to manage effectively was first, the knowledge of the task (job) which they were expected to perform, and second, the ability to exercise the authority invested in the managerial position to get the job done. Managers used their authority to direct and supervise their subordinates towards the achievement of the stated goals and objectives.

Since the authority which was needed to execute the plans, by and large, depended on the position of the incumbent within the hierarchy, issues such as seniority, the nature of the job (project), length of service and the degree of access to resources, in particular access to and control over the financial resources of the enterprise, contributed most to the managers' status and power. By being in possession of job-related knowledge, power and authority managers were assumed to be able to get the job done.

Project managers were no exception to this rule. Their ability to manage project organizations and their managerial potential for successfully completing a project was often judged and decided upon through their past work experiences, the size and nature of the project for which they were responsible, its budget, and the number of subordinates under their direct control.

How the job was done, legitimate or otherwise, and how control over the people and work processes had been gained and maintained were of no real consequence. Very little attention was paid to how people behaved and interacted with one another in the work environment. Managers were expected to concern themselves with those aspects of people's interactions at work which were seen as vital for the ongoing operation of the mechanically structured organizational activities. In charge of a machine-like organization, the managers required very few of the modern managerial skills to remain in control.

As the concern for the development of organizations gained momentum, the role of managers also gained importance, but the managers' own development did not, as yet, constitute a major issue, nor did its connection with the overall development of the organization.

Functional skills

The start of a more systematic search for the discovery of what skills managers need is probably signposted by the work of Henry Fayol (1916), a French engineer, who attempted to identify what managers had to do and their function in the maintenance of the organization. The functions of managers described in Chapter 3, that is those of analysing, planning, organizing, monitoring, authorizing, and enabling, are based on his ideas. Fayol added communication as a function which provides the link between the other functions. But communication was defined with the emphasis very much on the availability and use of information and not necessarily the mastery of interpersonal skills displayed by modern managers. In the context of the management of projects, emphasis was also placed on the need for the project manager to be familiar with formal communication, top down (one way) and mostly of a written nature (memos and progress and final reports).

Another interesting aspect of Fayol's work was his consideration of the position of the manager within the hierarchy in the organization and the degree of need for the kinds of skills described earlier.

It was considered that while operatives required technical and solely task-related skills and knowledge, managers who occupied executive positions needed skills of managing work (determining objectives, forecasting, planning and organizing) and managing people (directing, co-ordinating, controlling and communicating).

The assumption underpinning the managerial skills described above was the Model of Man used by theorists half a century ago to describe the individual who worked in work organizations in the position of an subordinate employee (Barnard 1938; Blue 1955). This was based on mechanical presuppositions, or at best, a simple biological organism which required constant direction and had to be controlled on a continual basis. In short, people in work organizations were thought of primarily as rational entities functioning in a predictable manner. People were not considered as being what they really are – who behave, respond and interact with others and their environment. Their need and desire for independence and the exercising of autonomy and control over their work were, by and large, neglected.

Task-related skills

Frederick Taylor (1911), the father of scientific management, carried out his studies and experiments at the Bethlehem Steel Company in the USA, (referred to in Chapter 3). As an engineer, he was somewhat preoccupied with 'objectivity' and 'order' and the need for exercising control over people and organizations as though they were machines or parts of it. This was achieved largely through paying attention to the ways in which operatives carried out their tasks. He strongly believed that managers needed the technical skills for design and measurement of the job in order to undertake the responsibility of ensuring that the task was carried out in the most efficient and economic manner. This required breaking down the operative's job into basic and manageable components. Taylor then attempted to find the best way (method) of doing the job.

His preoccupation with the measurement of work inevitably called for managers to possess both sets of skills suggested by Fayol, namely 'managing work' and 'managing people'. They also needed skills to be able to use effectively the authority vested in them as a result of their position within the organization, in order that maximization of productivity is assured.

Of course, the followers of Scientific Management showed concern for the behavioural aspects of people in the organization, but they were mainly criticized for adopting a rather simplistic model for managing people at large. Since people in the organization were generally believed to be primarily motivated by financial incentives, managers had to show that they were able to motivate people by offering reward (money) and using coercive measures (punishment), in order to achieve the maximum output from a given input. The necessary skills for managers were thought to be job design and measurement, finding the best way of getting the job done in the most effective and least costly way possible.

McGregor (1961) describes the managers who strictly follow the principles of Scientific Management as those who believe in 'Theory X'. In his view, Theory X managers approach people with the fundamental assumption that people are basically lazy and therefore do not enjoy work or responsibility. Managerial control, therefore, is needed to ensure that people do as they are told. The motivational technique used by such managers, McGregor suggests, is either the use of 'seduction' through providing financial rewards or 'coercion' and the use of fear to ensure obedience. A fuller discussion of McGregor's theory is included in Chapter 10.

Information-processing skills

The advocators of the Weber, Fayol and Taylor philosophy all share a belief central to their attitude and philosophy to work and people. They regarded work organizations as operating in a closed system with little interaction between its wider environment. The product or services were regarded as the outcome of the internal processes of transformation of input resources, such as technology, people and managerial skills. These were believed to provide the essential input for almost any work organization.

Communication was also regarded as an important aspect of the management of work organizations and naturally managers used this medium, either in spoken or written forms, to ensure the efficient transformation of the resources into products and services.

Later studies concentrated on the role of the manager as an 'information processor' who, as Kakabadse et al (1987) note, is primarily involved in 'collecting information, filtering and monitoring and using the data to organise and control the people and their activities. Managers were held accountable for the smooth operation of the information system which in turn improved the performance of the people involved'.

Managers process the information primarily to make decisions. The advent of computers enabled managers to analyse large amounts of data and use them to achieve the standard, quality and quantity which their organization was aiming for.

In the field of operational and management research, the importance of processing information available to managers for the purpose of effective decision making constitutes the central theme. The application of management information system (MIS) techniques and the availability of personal computers (PCs) has made the process of dealing with and effectively using data to manage tasks and people much easier (see Chapter 9). Part of the process of managing people in an organization is that of providing the end users of the information with relevant data in form of feedback. It is expected that managers will do this skilfully and on time.

For example Carnall and Maxwell (1988) cited an example of a water authority which, by introducing computer control of water flows and pressures, not only increased the accuracy of usage and production, but also made major savings by optimizing electricity consumption, ensuring minimized water wastage and reducing personnel and maintenance costs, all this through having a precise knowledge of users' needs.

Today's project managers have benefited from the advancement in technology in the field of micro-computers and the availability of project planning software. This has enabled them to implement complicated multi-stage project plans, knowing at any given stage the exact state of project progress and any possible shortcomings in the future. To be able to do this, project managers are expected to handle data skilfully, in that they should know where to find the data required, how to analyse them and how to interpret the available information expertly to their own advantage.

Project managers frequently find themselves needing to make decisions. Being in charge of a project brings with it the necessity to make decisions quickly and accurately. The skill of processing data is, therefore, one of the most essential tools in the project manager's armoury.

Mintzberg (1973), among others, identifies decisional roles. The entrepreneurial, disturbance handler, resource allocator and finally negotiator explain not only the need for managers to be in possession of accurate and up-to-date information, but also highlight the issue that managers are expected to make correct decisions on many aspects of work within the organization.

The entrepreneurial aspects of managers' work requires them to make decisions concerning the future of the organization or project in order to make it more effective. How the objectives of the organization should be modified, what kind of product should be developed for the future and what kind of technology will be needed to satisfy the task-related aspect of the organization, are all questions that require decisions to be taken.

The role of managers as disturbance handlers requires them to make short-term decisions, which may have serious consequences for the future effectiveness of the organization. An important aspect of managers' jobs is also that of deciding on the resources required at any point in time and how they should be utilized and deployed on tasks throughout the organization. Managers decide who gets what and work out how much of their own time is going to be spent on each task and decision. Vroom and Yetton (1973), while relating the managerial style to the degree of participation on the part of subordinates in managerial decision making, argue that the following are the next important components of managerial decision-making

1. the quality of the decision to be made
2. the degree of acceptability of the decision by the subordinates
3. the time taken to complete a decision-making process.

What, however, remains undisputed is the fact that managers require quality information for prompt decision-making. Based on the complexity of the issue and the urgency for the decision making, managers must

1. gather as much relevant information about issues on the situation in hand
2. assess the quality of the information available to them
3. seek the views of those who will be affected by the decision
4. list alternative ways in which the problematic issue or situation can be satisfactorily resolved

5. get feedback by assessing the consequences of the decision and incorporate the lessons learnt in future decision-making situations.

Managers need to develop the ability to determine how much of the information made available is needed for accurate decision-making and most important of all the exact time for decision-making. An effective manager seeks the relevant information and makes decisions at the right time.

People Skills

Mayo (1945) and a colleague from Harvard University, while investigating workers' responses to external stimuli and how they can be improved, came across a phenomenon which at first they found difficult to explain. The mysterious phenomenon was later explained as having something to do with the social network of relationships which people create in an organization while accomplishing a given task. A whole school of behavioural research took off from this point, which had implications for training and educating managers working with people.

Unlike the followers of Classical and Scientific Management, who placed the 'task' in the centre of their analysis, the advocators of the 'Human Relations School' argued that 'people' make up one of the most important resources for any work organization. The effective management of people enables managers to achieve the organization's objectives more efficiently and this requires effort, knowledge and skills on their part to understand, motivate and lead people. Managers need an interest in optimizing the benefits from the effective utilization of the human and other resources which are available to the organization.

Managing people, as explained in detail (see Chapter 10), requires the individual, or the team of individuals in charge of the project, to possess other skills besides designing jobs, measuring, monitoring and evaluating output. Rather, it is argued by followers of the Human Relation Schools that managers can alter and increase the effectiveness of their personnel, and ultimately the efficiency of the organization, by managing people skilfully. This calls for a new range of interactive and interpersonal skills for managers, such as effective listening, communication, problem-solving, decision-making and handling conflict at work.

McGregor (1961) described this new approach to managing organizations as 'adhering to theory Y principles'. The implications were that managers must recognize that people enjoy work and that financial reward and punishment are only part of the package available to managers for getting the most out of their workforce.

Subsequently, motivating people became an essential skill for managers and, in most situations, replaced the need to resort to punishment. Concepts such as 'management by walk-about', people's active participation in decision-making as well as their involvement in executing those plans, taking part in improving the quality of work and life on a collective basis, became the focus for people managers.

Role-related skills

Mintzberg (1973), in his study of managers, posed the question 'What do managers actually do during the course of their daily work?' His observations of the activities that managers were actually engaged in revealed that during the course of their daily activities they found themselves in many varied situations, each of which required them to call upon specific skills in order to carry out certain roles effectively. Managers are often spokesmen on behalf of the organization and the project which they are responsible for in one situation and take on the responsibility for being resource allocator or negotiator in others. These responsibilities are described as 'roles' by Mintzberg and are discussed at some length in Chapter 3. They have become a powerful analogy which is frequently used to describe the constituent components of the managerial job, in order to play each set of roles proficiently the possession of certain role-related skills is paramount. Managers, according to Mintzberg, cannot exercise their authority unless they are enabled to play all three categories of roles namely interpersonal (leader), informational (diplomat) and decisional roles (chief executive). The implications of Mintzberg's study for management educators and trainers are profound and far reaching.

The range of skills which managers require is extensive. Managers need 'soft skills' such as those which enable them to deal with people as well as the 'hard skills' related to the nature of the task in hand. Unfortunately, as Kakabadse et al (1987) argue, it is often the decisional aspects of managers' jobs which receive far more emphasis than the interpersonal ones. It might be argued that this is the wrong way round.

The contemporary situation

What constitutes essential managerial skills has been the subject of frequent discussion and change. In each era one aspect of the manager's job has been the focus for attention and has subsequently been emphasized.

With the advent of Scientific Management, it was the task and its related-skills, which received the most attention. From the Human Relations perspective, people were thought to be the main component of the organization, and therefore people-related skills were widely prescribed.

Expectations of modern managers' ability to perform varied roles within their organizations are influenced by the presence of two major issues; first, that managerial work is a complex and multi-dimensional profession, and second, that with the advancements made in various fields of inquiry (as well as increased communication and the fluctuation in the socio-economic climate) a better understanding has been gained of the nature of managerial work and the kinds of skills needed by managers to do their jobs effectively. Yet, to the practitioners, these developments in the field of management may not appear entirely clear cut.

Managers in industry, especially in industrialized countries, are believed to have different sets of needs than those of their counterparts who work in developing countries. Managers who are placed in charge of development projects in developing countries are bound to experience different sets of expectations from their colleagues, peers, subordinates and even their superiors, to those who manage projects within their own technologically advanced societies.

The available textbooks on project management are generally unhelpful in so far as managerial skills for development of the project managers are concerned. Often the task-related skills have been over-emphasized at the expense of other skills. Furthermore empirically sound attempts specifically aimed at determining the management development needs of project managers in developing countries are scarce and first-hand data are generally hard to come across.

Most texts on project management tend to list the ideal and not necessarily the desirable attributes and characteristics which it is expected a project manager will possess. Such attempts are too often intended to seek out the 'best manager' rather than being used as a guide for project managers to help themselves in their own managerial development and the development of their institution.

An integrated framework for the development of managers is needed, which could be adjusted and used by practitioners regardless of the nature of the job or size of the project, the executing organization concerned or even the cultural and other socio-economic differences which are present in various developing countries.

Towards an integrated model of management development skills

An attempt at this is described in the rest of this chapter and is largely based on two studies carried out by Analoui (1989, 1991). The first study was conducted in 1989 in the Zimbabwean Public Service and involved sixteen organizations. These included ministries, parastatals and research institutions. The second study was carried out in 1991 and 1992, it involved middle and senior managers who worked for the Indian National Railway.

The result of these two studies and the data generated from interviews has provided the basis for the construction of the Integrated Model of Managerial Skills needed by project managers (see Figure 5.1).

The studies suggested that, though different project managers and officials showed different preferences and specified different priorities in so far as their perceived needs for managerial skills were concerned, the analysis of the data revealed that they tended to share the need for only three broad categories of managerial skills and knowledge as opposed to the five considered earlier in the Chapter. These three identified categories of managerial skills provided the three major components of this integrated model. They are

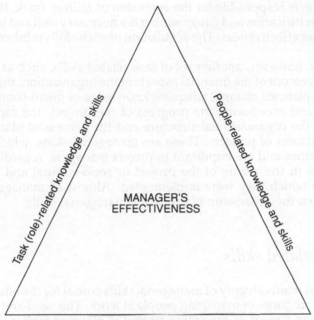

Fig. 5.1   An integrated model of managerial skills

1.   task-related skills
2.   people-related skills
3.   self-development skills.

Managers and senior officials can be effective only if they have benefited from a balance of the above skills. To acquire one set of skills without possessing any of the others will lead to partial development and thus partial effectiveness of the managers concerned.

In order to signify the importance of the above for the development of effective management, each component of the above framework of analysis will be further explained in some detail. It will then become apparent that managers have to break out of the traditional mould which places stress solely on the task-related aspects of their job, and consequently expect them to possess only task-related knowledge and skills. Managers need to view their own development and their organization's effectiveness within an integrated context.

Task-related skills

The studies carried out by Analoui in southern Africa and India show that managers, especially senior project managers, require task-related skills and knowledge. Task-related managerial skills are those which are specific to the nature of the task to be performed. For example, for a senior official who is responsible for the extension of railway track, the knowledge of electrification and gauge setting is a necessary skill and is essential to his or her effectiveness. The acquisition of such skills is inherent in the job itself.

There is, however, another set of task-related skills, such as the effective management of the financial aspects of the organization, the ability to introduce planned change, adequate knowledge of micro-computers for planning and monitoring the progress of the project, the capability of analysing the organizational structure and the successful planning and implementation of projects. These are managerial skills, which are of a general nature and are important to project managers, regardless of the differences in the nature of the project or socio-cultural and economic context in which they were implemented. Almost all managers would benefit from the possession of the above managerial skills.

People-related skills

The second main category of managerial skills crucial for the effectivess of managers are those of managing people at work. The need and necessity for acquiring the art of managing people is often understated and has traditionally been neglected.

The fact that this category of managerial knowledge and skills has been generally taken for granted was particularly acknowledged by senior managers involved in these two studies, who had first-hand experience of both implementing projects as well as managing administrative tasks at middle and senior levels.

Managerial skills such as effective communication, problem-solving, handling conflict at work, decision-making, motivating and leading and developing teams (see Chapter 10) are essential for project managers. Although the means and ways by which people communicate with one another differ from one social setting or organizational context to another, the results of these studies show that the need for listening skills, effective presentation and other interpersonal skills is shared by all managers. This is simply because managers have to rely on people to get the job done. For instance, most of the managers who were involved in the surveys believed that skills such as managing negotiation processes and effective communication, whether at individual, group, interdepartmental, ministerial or even international levels, are essential for the smooth operation of any organization.

Managers who benefited from a balance of task and people-related skills found that they managed projects with more ease than those who specialized only in specific task-related skills and knowledge.

The traditional approach to the management of work organizations, as a whole, and project management in particular, has tended to over-emphasize the technical, task-related experiences, knowledge and skills needed by managers and thus to undervalue the less tangible socio-psychological skills necessary for dealing with people at work.

Self-development skills

The third and probably the most important category of managerial skills, necessary for the development of all managers, is that of self-development. This aspect of management development has hitherto been the subject of considerable neglect. Today's managers are making decisive moves towards self-development and the need for acquiring basic skills as a means of recognizing their own abilities and strengths, as well as identifying weaknesses and potential areas of future development.

The traditional approach to development of management placed the emphasis largely on what skills and knowledge managers require and how best such skills should be developed. They were not encouraged to realize who they were, to identify the level of personal and professional needs, and most importantly what means and strategies were available for them to realize their desired level of development and hence their performance at work. In today's modern organizations the development of managers is viewed as part and parcel of the organization's total development. The responsibility for the managers' development has been partly given to managers themselves, while the organization ensures that they are provided with the ability, that is the relevant skills and knowledge and the opportunity to achieve their personal/organization objectives. Management educators and development specialists nowadays utilize a series of self-inventory psychological tests which not only enable them to assist managers in their quest for self-development, but also provide managers with initial information concerning their level of intelligence, aptitude, preference for styles of leadership, preferred approaches to deal with conflict and a host of other relevant information which enables managers to achieve self-awareness. Once managers have obtained the initial information about themselves and how they interact with the world, whether workplace or otherwise, in which they operate, they will then be able to proceed with the task of self-development on a planned and if necessary supervised basis. What must be remembered is that managers' jobs constitute the most important aspects of their career and development. Therefore, it is of utmost importance that they are enabled to identify the levels of task and people related skills required for remaining effective and at the same time provide them with an opportunity to gain the necessary knowledge, skills, values and attitudes required.

The Integrated Model of Managerial Skills (Figure 5.1) is based on the thesis that there is an undeniable link between individual and organizational development. This micro-model for developing the organization views the individual employee as an integral part of the work organization and, therefore, the development organization's human resources is expected to contribute to improved productivity and effectiveness.

The hierarchy of needs for managerial skills

One of the most intriguing discoveries made during the course of the research concerned the extent to which identified categories of skills were perceived as essential by managers who functioned at junior, middle and senior levels of hierarchy.

What emerged from the analysis of the data was that there seemed to be a hierarchy in operation in so far as the needs for managerial skills were concerned. Moreover, an inverse relationship between the designated role of individuals and their perceived need for acquisition of task-related people-related and self-development skills was discovered (see Figure 5.2).

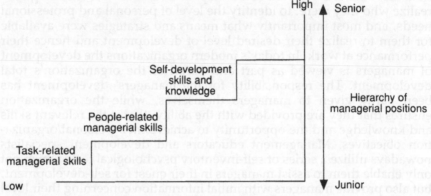

Fig. 5.2 Ascending hierarchy of managerial skills

It must be also noted that the formal position of individual project managers in the organization and their seniority seemed to act as a determinant factor when their superiors were deciding on the size, nature, degree of complexity and even the number of projects with which they could be entrusted.

The general conclusions reached are first, the lower the position of the individual manager in the organization the higher the need for task-related skills. Second, the higher the position of the individual manager in the hierarchical order of formal status in the organization, the more likely he or she will be in need of people related managerial skills. Senior managers, who already possess substantial managerial knowledge and skills for dealing with task and people, tend to strive towards self-development.

Managers appear to require a balance of people-related and task-related skills on one hand and knowledge of self-development on the other. It is often the degree of realization of the latter which tends to facilitate or inhibit the effectiveness of the use of the former skills which are necessary for the successful implementation of projects.

The modern approach to the development of managers specifies closer involvement by individuals in the process of identifying, designing and putting into practice their own management development programmes.

However, this can work satisfactorily, only if managers are aware of the extent of their own need for managerial skills. Self-development begins with a conscious attempt on the part of managers to determine the gap between their present performance, knowledge and skills and the desired level expected from them or for which they are aiming. Since the acquisition of managerial skills and knowledge is a lifelong process, managers need to be prepared to take the responsibility for their own development.

How do managers learn?

So far the discussions in this chapter have evolved around what kinds of skills and knowledge managers require to become and remain effective. However, in order to realize these objectives managers need to know how they learn. This aspect of learning constitutes an important aspect of a manager's self-discovery and development.

The orthodox approach to educating and training managers tends to concentrate on the task-related knowledge and skills, but has also been built on the presupposition that managers ought to be taught in a structured fashion and controlled environment. For example, in order to implement a typical management-training programme, the training needs of individual managers are identified (assumed), a programme of management training is chosen, and managers are then expected to acquire the kind of knowledge and skills which are 'provided' for them. In the role of 'passive' learners, managers are, more often than not, exposed to a series of lectures, case studies, role-play exercises and similar activities without consideration being given to their preferred learning styles.

The issue of self-development, discussed earlier as a part of an integrated approach to management development, also includes the knowledge of one's own attitude, orientation and preference towards 'learning' styles and methods. Nowadays, it is believed that sustainable management development necessitates managers being trusted with the responsibility for their own learning. Unfortunately, the inability to learn effectively and the lack of relevant knowledge concerning the way in which their individual learning takes place, can result in a deep sense of dissatisfaction and may even lead to the ineffective acquisition of managerial skills.

Harvey and Mumford (1982), argue that managers have to 'learn to learn'. It is suggested that most managers are unaware of the ways in which they can learn most effectively.

Learning styles

Kolb, Rubin and McIntyre (1974) developed the first 'learning style inventory'. This method categorizes managers into four types – the converger, diverger, assimilator and accommodator. These classifications are primarily based on how managers learn from 'concrete experience', how 'reflections and observations' result in new learning, how the managerial 'concepts' learnt are then 'formulated and generalized' in other situations and finally, how managers go about testing the acquired concepts in new situations.

Kolb's original work was later developed further by Harvey and Mumford (1982) and is used by management educators and developers to assist managers in discovering their most preferred style of learning. In this way managers can learn about their own approach to the acquisition of managerial knowledge, skills, values and attitudes and thus will carry on learning after formal management training is completed.

Harvey and Mumford (1982) identified four different approaches to learning: activists, reflectors, theorists and pragmatists.

Activists enjoy the challenge of a new experience. They are sociable and tend to 'throw caution to the wind'. They often use 'trial and error' as a basis for learning new experiences, yet tend not to carry forward their acquired experiences into the 'doing' stage.

Reflectors prefer to consider issues from different angles and think about the varied implications of an issue before making the final decision. They often take a back seat while observing others to make sure nothing is missed out in their analysis before arriving at a conclusion. The caution with which reflectors approach issues and decisions indicates that they like to 'sleep on it' before they act.

Theorists like to put their own experiences and observations together to arrive at a sound and rational theory. They make models in which they place events in a logical manner. The process of decision making for theorists is generally based on subjective rather than objective reasoning. The desire for certainty and control over circumstances makes them wary of those who are judgemental or rely on their intuition tending to adopt unorthodox ways of doing things.

Pragmatists are attracted to sound ideas and put them into practice if they are attractive enough or if they work. The desire to solve problems and generally get on' with it means that they do not like wasting time and tend to react to situations rather quickly. Pragmatists like to test theories and new ways of doing things. They like to make practical decisions.

To highlight the importance of the above, observe a situation where two individual managers, one a theorist and the other a pragmatist, are exposed to a lecture on motivation and productivity. The theorist is unlikely to put what is learnt into practice unless it fits into his/her already established schema. However, the pragmatist is prepared to adapt and apply the principles learned as long as they are attractive enough to her/his and get the job done quickly.

As indicated earlier, an important aspect of self-development for managers is gaining substantial understanding concerning themselves – their strengths and weaknesses. This knowledge will provide the foundation for future development. The question that has to be asked is 'what is my preferred style of learning and how best can I improve on my existing task, people and self-development knowledge and skills?' Improving managerial skills and knowledge is a lifelong process which has to be approached on a planned basis.

Conclusion

Effective management requires different skills, knowledge and understanding. The significance of each identified category of managerial skills has been judged differently in different eras. At the beginning of the twentieth century skills of a technical nature together with the ability to deal with the task in hand were considered crucial for the management of organizations and projects. Since then, socio-psychological skills for dealing with people, the effective use of information and finally the ability of managers to act as the performers of many different, but interrelated, organizational roles has become the main focus of attention.

Having an integrated model of management skills helps managers assess their own professional needs for relevant managerial skills in the context of their own specific situation. Such a realistic approach to management development places emphasis on the managers as the central figures in relation to the people with whom they work and situations in their immediate work environment and beyond.

References and further reading

Analoui F 1989 Project manager's role: towards a descriptive approach. *Project Appraisal* **4** (1).

Analoui F 1990 Effective management with people skills. *Journal of Managerial Psychology* **5** (3) June: i–iii.

Analoui F 1991 Project management in context of change. In C Kirkpatrick (ed.) *Project rehabilitation in developing countries.* London, Routledge: ch. 12.

Barnard C I 1938 *The function of the executive.* Cambridge, Mass, Harvard University Press.

Blue, P M 1955 *The dynamics of bureaucracy.* Chicago, Chicago University Press.

Carnall C, S Maxwell 1988 *Management: principles and policy.* Cambridge, Institute of Chartered Secretaries & Administrators.

Cooper G 1981 *Psychology and management.* New York, Macmillan.

Fayol F 1916 *Administrations industrielle et generale.* Paris, Dunod.

Fayol F 1949 *General and industrial management.* London, Pitman.

Harvey P, A Mumford 1982, *The manual of learning styles.* Berkshire, UK, Honey.

Kakabadse A 1983 *Politics of management.* Aldershot, Gower.

Kakabadse A, R Ludlam, S Vennicombe 1987 *Working in organisations.* Aldershot, Gower.

Kanter R 1983 *The charge masters.* London, George Allen & Unwin.

Kolb D A, L M Rubin, J M McIntyre 1974 *Organisational psychology: an experimental approach.* Englewood Cliff, NJ, Prentice-Hall.

McGregor D 1961 *The human side of the enterprise.* New York, McGraw-Hill.

Maugham L 1979 *The politics of organisational change.* New York, Associated Business Press.

Mayo E 1945 *The social problems of an industrial civilisation.* Cambridge, Mass., Harvard University Press.

Mintzberg H H 1973 *The nature of managerial work.* New York, Harper & Row.

Taylor F 1911 *The principles and methods of scientific management.* New York, Harper & Row.

Vroom V H 1964 *Work and motivation* New York, Wiley.

Vroom V H, P Yetton 1973 *Leadership and decision making.* University of Pittsburgh Press.

CHAPTER 6

Project implementation planning

Introduction

This chapter is concerned with one of the primary tasks of project management, that of preparing for project implementation. The definition of projects in this book includes reference to the fact that they are investments which have definite start and end dates. The process between the start and end dates is that of project implementation or establishment. This process involves a series of activities which need to be planned, operated and controlled and which will inevitably involve the utilization of resources. The management of these activities so that the project can be completed on time and at cost consistent with the plan is a fundamental requirement of project managers.

In Chapter 2 the 'project framework' was described as a tool for planning, monitoring and evaluating a project. In this chapter the project framework is used as the basis for implementation planning. The implementation process can be described as the process whereby 'project inputs are converted to project outputs as set out in the project framework'.

This simple definition, however, conceals some complications that can arise with different types of projects. In Chapter 1 a broad distinction was made between capital-intensive and people-based projects, and these were linked, respectively, with the blueprint and process approach to project development. In the case of capital-intensive projects the implementation phase is normally easily identified and has a readily defined starting-point and end date when constructed facilities can begin operation. A useful analogy is that of building a ship: the project phase is the building of the ship and the operational phase is that of sailing the ship between ports.

People-based projects do not fit easily into this pattern and their project or establishment phase is more complicated. Examples of such projects are agricultural and rural development projects and those involved with community development. With these projects the establishment and operation phases overlap, thus creating the opportunity to substantially 'redesign' the project during its establishment, the basis of the 'process approach' to project implementation.

This is particularly important for people-based projects in which a local population is expected to participate fully in project activities. Such projects are much less suited to the techniques of implementation planning than capital-intensive projects for which detailed plans must be prepared before implementation can begin. The assumptions that need to be made for planning implementation of process-type projects may be so large that in practice the schedule for implementation becomes little more than wishful thinking.

The point of issue here is that managers must recognize the value and limitations of implementation planning in relation to their particular project and its environment. Many of the techniques discussed in this chapter have been adapted from those used for major technologically based projects in western countries. It is, therefore, legitimate to question how useful these techniques are for people-based projects in countries that have chronic problems such as severe domestic budgetary constraints, political and social instability and poor communications systems. On the other hand it might be concluded, albeit with reservations, that such techniques are of some significant practical use. To be effective, managers must have a systematic basis for organizing the activities that collectively result in the completion of a project. Implementation planning is the basis of so much else regarding the project process. It is the basis for budgeting for money and resources, the basis for identifying bottlenecks and testing assumptions, and the basis for accountability and measurement.

The sequence of implementation planning

As previously stated the project framework is a useful starting-point for planning the implementation of the project. The project outputs as detailed in the narrative summary indicate the components of the project which are to be completed. For projects which do not have a prepared project framework an initial task for managers might well be the drafting of a framework along the lines already discussed in Chapter 2.

Knowledge of the project outputs then makes possible the breaking down the project into 'manageable' components. This is generally referred to as the work breakdown structure (WBS). The definition of manageable components is elastic, but may be broadly defined as an activity or series of activities that can be carried out independently from other elements of the project. Typical examples might be as follows:

Activity		Project
• infrastructural development	}	Integrated rural
• farmer training		development project
• factory construction		Sugar
• plant installation	}	plantation
• land development		project

Clearly each of these components, while being part of the same project, are for most purposes entirely separate from one another.

Preparing a WBS has a number of uses, not all of which are related to implementation planning. As briefly mentioned in Chapter 4, one of these is the setting up of the project organization. It would be difficult to establish a project organizational structure without first breaking down the project into its manageable components and subcomponents. The process can also lead to the identification of more accurate staffing and skill levels required for executing different aspects of the project. A major task for managers at the start of a project is identifying skill requirements and appropriate staff to carry out specific tasks. Establishing a project organizational structure following the systematic breaking down of the project into its component parts can be most helpful in this respect.

Another major purpose of breaking down the project into manageable components is to more clearly identify those individual activities that, collectively carried out, ensure the completion of the project. Subsequently four primary factors can be determined for each individual activity, each activity being a task which can be carried out independently from other project activities. These are

1. the sequence of project activities
2. the resource requirement for undertaking the activity
3. the estimated time it will take to complete
4. the responsibility for undertaking it.

Once these factors have been determined, managers have a firm basis for establishing an implementation schedule, a resource budgeting system and an organizational structure within which project staff are aware of their individual tasks and responsibilities.

This also provides a basis for a system for monitoring project progress, resource use and staff performance. Often termed a 'management information system', such a system will depend on the nature, size and complexity of the project and its organizational structure. An important point is that a management information system is not something that is worked out independently once the project is underway but ideally developed beforehand as an integral part of the implementation planning process.

A summary scheme of the implementation planning process is shown in Figure 6.1.

The left-hand column of Figure 6.1 is covered in Chapter 4 and management information systems in Chapter 9. The focus of this chapter is the activities in the right-hand column of Figure 6.1. While much of the remainder of the chapter will be devoted to the techniques of implementation planning, it is essential that the importance of managers' knowledge and judgement is not overlooked in this process. It is too easy to make the implicit assumption that planners' and managers' knowledge readily encompasses such factors as, for example, the time it takes to complete an activity or what the logical sequence of a series of activities is. In fact such knowledge is not easily acquired and is subject to a large

degree of uncertainty and error. For some projects therefore, the application of forecasting methods and statistical probability analysis are complementary tools for use in undertaking implementation planning.
Coverage of these more advanced techniques of implementation planning is beyond the scope of this text and readers are referred to the books
listed at the end of the chapter.

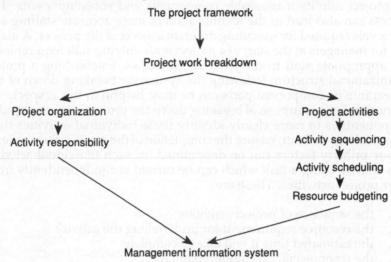

Fig. 6.1 Project implementation planning schema

Project framework and the work breakdown structure

The project framework has been extensively discussed in Chapter 2. The
principal role that the framework can play in planning for project implementation is that of providing information on the proposed project outputs. It is the project outputs that are the end product of the
implementation process.

The outputs will be ultimately operated to achieve the project objectives but for implementation purposes it is the achievement of the outputs
themselves that are the objectives.

As indicated earlier the project outputs provide a basis for the process
of breaking the project down into its manageable components through
the work breakdown structure. These can be identified at various levels
from relatively large components which are the major outputs of the
project down through further levels until a component becomes an activity which can be implemented more or less independently of any other
activity. This is best understood through an illustrative example.

The example project is an outgrower and estate sugar cane production
and processing project, which involves the establishment of a cane nurs-
ery and plantation and the construction of a facility for cane processing.

For convenience it is easy to consider level 1 of the breakdown structure as simply The Project itself. Level 2 indicates the main components of the project in terms of its principal outputs, that is the capacity to produce sugar cane in the plantation and the capacity to process the cane in the factory. But this does not include any element of overall management that, in itself, is a major output for most projects. Therefore, the management unit (MU) might well be considered a level 2 component.

There are no hard and fast rules on carrying out a work breakdown and some judgement is called for. Generally the level 2 breakdown would consist of the MU and the major project components.

At level 3 each of the major components is further divided into its subcomponents. In this project the level 3 breakdown might be as shown in Figure 6.2

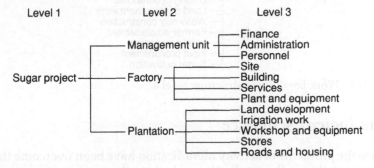

Fig. 6.2 Work breakdown structure: sugar plantation project

At level 3 the project subcomponents are clearly identified. While there is a dependency between all the components in terms of the overall project each component might be considered separately for the purpose of resource allocation, planning, administration and staffing.

The combination of levels 2 and 3 may form the basis of the project organizational structure. This should be considered in conjunction with the discussion of organizations in Chapter 4.

At the fourth level of breakdown it is probable, but not necessarily the case, that the project elements are no longer project outputs or subcomponents of outputs but identifiable activities.

'Activities' in this context have a particular meaning. They are defined as tasks requiring resources and time for their completion and which are independent of other tasks continuing concurrently. They are tasks that could proceed to completion even if all other project activities ceased. This is not to say that activities are totally independent from each other. But the dependence is in terms of what activity, or group of activities, must be completed before the activity in question can actually begin, and what other activities can begin only after the activity in question is completed. Activities must also be relatively homogeneous in that they involve use of resources in a generally consistent manner over the duration of the activity implementation.

Using the sugar project example, the work breakdown structure can lead to the identification of 'activities' for the Land Development component as shown in Figure 6.3.

It can be seen that there are several 'activities' which need to be carried out before the land development component of this project is completed, ending with the cultivation of cane. The level 4 breakdown in this case yields identifiable tasks that need to be undertaken.

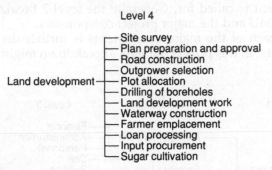

Level 4

Land development
— Site survey
— Plan preparation and approval
— Road construction
— Outgrower selection
— Plot allocation
— Drilling of boreholes
— Land development work
— Waterway construction
— Farmer emplacement
— Loan processing
— Input procurement
— Sugar cultivation

Fig. 6.3 Work breakdown structure: level 4

The logical sequence

Once the problems of activity identification have been overcome the next important step is the application of logic to the sequencing of the activities. This is the step that is the most important of all and it requires good judgement, knowledge and experience. It might be argued that this is what a project manager is paid to exhibit and why a successful manager is generally well remunerated. The old adage of good managers being born not made is worth recalling. It does sometimes appear as if some managers are instinctively able to programme activities so that things are achieved and targets met without any seeming effort.

Undoubtedly people have different capacities for carrying information in their heads and for seeing things in a more structured way than others. Such people have an advantage in this area of management. However, the larger and more complex the project, the greater the need for a more systematic approach to planning activities.

This part of the implementation planning process requires the definition of the logical sequence in which the activities can be undertaken. Certain activities must be completed before others can begin and sometimes an activity can begin only after several activities have been completed. This is obvious to anyone who has ever cooked a meal or even made a cup of tea. Before heating the kettle it must be filled with water and the tea can be made only once the water has been boiled and added to the leaves in the teapot. There is a logical sequence to these activities which cannot be rearranged. To drink the tea it is necessary to set out cups and saucers, make available milk and sugar and possibly even lay all these out on a tray. These activities also follow a logical sequence: it is not usual to add milk to the cup before the cup is taken out of the cupboard! However, the activities associated with preparing the cups and saucers are not logically linked to the activities associated with boiling the kettle and can therefore take place simultaneously. Of course to drink tea the two sets of activities eventually must come together through some linking activity, such as pouring tea from the teapot into the cup. If this activity was missing there would be no logic to the whole process. There are some pitfalls to watch out for in deriving an activity sequence.

For example a difficulty occurs with activities that, although dependent on one another in a sequential way, do not require the completion of one before the other can begin. In the case of the sugar plantation project the drilling of boreholes and farmer settlement might be two such activities. Although farmer settlement cannot begin until some boreholes are drilled, the settlement operation is not dependent on the borehole drilling operation being fully completed before it can begin.

This presents a slight problem for the scheduling exercise. There are two ways to deal with it. The first is to recognize that there is an 'overlap' of activities with one beginning just ahead of the other. A more sophisticated technique, which can be used in certain types of this analysis allows for the identification of lag and lead times between the starts and finishes of the various activities. The second method is to break up the activities further, so that for example drilling boreholes is divided into two activities, one being 'initial drilling of boreholes' which is the amount needed before it is possible to begin farmer settlement and the other 'continued borehole drilling', being the overlapping activity of borehole drilling with farmer settlement.

The logic of the sequencing exercise is of crucial importance and, while valuable, techniques cannot substitute for the judgement required for the activity scheduling process. In the plantation example it is probably obvious that site survey is necessary prior to land clearance, but what about contour ridging and waterway construction? The answers to such questions need expert knowledge that will not necessarily be in the ambit of the manager.

It is therefore often prudent for managers to leave the detailed scheduling procedure to the heads of the different sections of the project organization and for the manager to concentrate on the more global scheduling exercise.

The techniques of network analysis

Once individual activities have been identified and a logical scheduling of the activities drawn up the result will be what is commonly referred to as a project network. This is simply a diagrammatic representation of the sequence of activities.

There are different ways of drawing networks. The commonest methods involve treating activities as arrows linking 'events' (the activity on line method), or treating activities as entities which are simply linked by arrows, but in which there are no events except the start and finish (the activity on node) method.

An illustration of networks drawn using the different methods is given in Figure 6.4. In each case the same set of activities and logical sequence are being depicted; for instance, activity *g* follows activity *c*, and activity *h* starts when activities *d* and *f* are both completed.

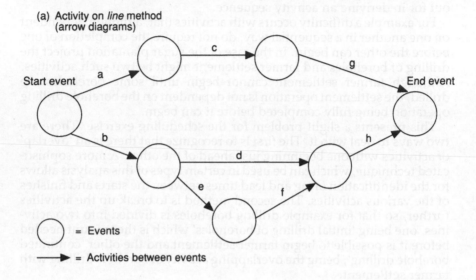

(a) Activity on *line* method
    (arrow diagrams)

Start event

End event

◯ = Events

──▶ = Activities between events

(b) Activity on *node* method

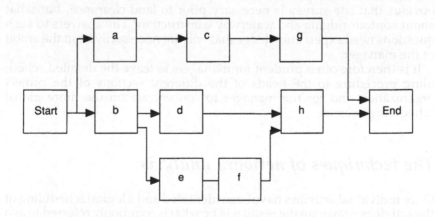

Start

End

[x] = Activity

──▶ = Signifies relationship between activities in terms of precedence

Fig. 6.4 Alternative methods of networking

There are no hard and fast rules to decide which is the better method to use but for this text the activity on node method, or the analysis bar charting (ABC) method as it is known, has been used for illustrative purposes. The method described below is mainly derived from Mulvaney's (1978) valuable text on the ABC method.

Networking has a number of purposes but perhaps the most significant one is that of providing the basis for determining the 'critical path' through the network by undertaking an analysis of the network. The critical path is the particular sequence of project activities which determines overall project duration. A major assumption underpinning this analysis is that certain information about each activity is known, in particular the *time* it will take to complete the activity. At the beginning of this chapter it was suggested that one of the main purposes of implementation planning was to determine the duration of project implementation. Timing of project implementation is of vital importance to projects almost by definition. In general the sooner an approved project is completed the better, particularly in terms of economic and financial viability. Also once a project is completed it releases management expertise to implement other projects.

By analysing the network it will be possible to identify those activities that actually determine overall project implementation time and thus give rise to the possibility of adjusting overall project completion time by completing those activities more quickly. The first step, however, is to determine duration time for each activity on the network. As with the logical sequencing of activities this process requires knowledge skill and judgement and will also be subject to a considerable degree of uncertainty.

Table 6.1 lists the activities associated with the implementation of the land development component of a sugar plantation project and indicates the logical sequence of activities. This is the first step to be undertaken in the implementation planning exercise. The second step is constructing the project network as indicated in Figure 6.5

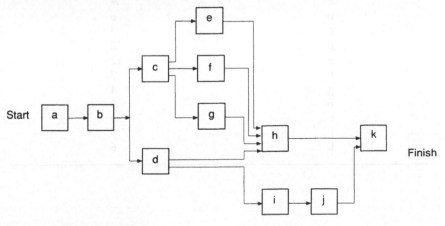

Fig. 6.5 Network diagram

Table 6.1 Land development component

Activities to be completed before activity can start		Activitity in question	Activities that can start only after activity in question has been completed
None	a	Resettlement area survey	b
a	b	Plan preparation and approval	c, d
b	c	Road repairs/construction	e, f, g
b	d	Settler selection/ plot allocation	h, i
c	e	Drill boreholes	h
c	f	Hand development work	h
c	g	Waterway construction	h
d,e,f,g	h	Farmer emplacement	k
d	i	Loan processing	j
i	j	Procurement of inputs	k
l,j	k	Arable plot cultivation	None

The network diagram is simply a diagrammatic representation of the information given in Table 6.1. Once the network has been constructed, which may take a number of attempts to sort out, it is possible to begin the analysis, provided the information with respect to activity duration has been derived. For this case study the activity duration times are given in Table 6.2.

Table 6.2 Activity duration

Activity	Duration in weeks
a	4
b	6
c	3
d	2
e	12
f	7
g	4
h	3
i	4
j	9
k	6

Network analysis: the forward pass

Assuming that the duration of each activity has been determined, and the network diagram drawn up, the next step is to begin the analysis by making a 'forward pass' through the network to determine the EARLIEST START and EARLIEST FINISH times for each activity and the project as a whole.

Using the activity on node method, it is helpful to use boxes to illustrate each activity and include the relevant times in the boxes as indicated in Figure 6.6a. The procedure is a straightforward iterative one. The EARLIEST START time for each activity is the same as the EARLIEST FINISH

(a) The forward pass
    (earliest start and finish times)

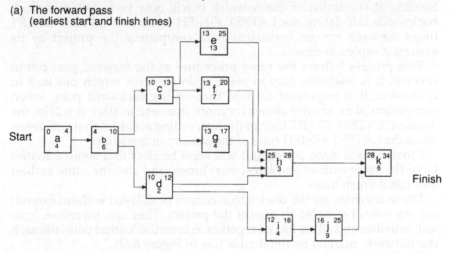

(b) The backward pass
    (latest start and finish times and
    the critical path)

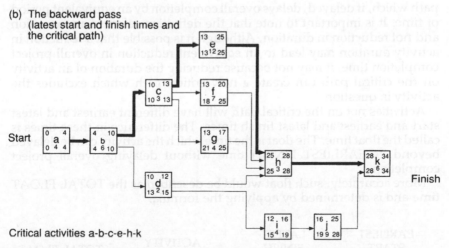

Critical activities a-b-c-e-h-k

Fig. 6.6 The forward pass, the backward pass and the critical path

time for the preceding activity. In the case of initial activities, the EARLI-EST START time is the same as the project start date. Where more than one activity is required to be completed before an activity can start it is the 'latest' EARLIEST FINISH time of the preceding activities that determines the EARLIEST START time of the activity in question.

The forward pass will provide the earliest completion date for the project which will be the EARLIEST FINISH time for the last activity to be completed.

The backward pass

Starting at the finish of the network it will now be possible to work backwards calculating the LATEST FINISH times and LATEST START times for each activity consistent with completing the project by its earliest completion time.

This process follows the same procedure as the forward pass but in reverse. It is relatively easy to make a simple error which can lead to confusion. It is important to note that on the backward pass, when completion of an activity allows for more than one to start, it will be the 'earliest' LATEST START time of the succeeding activities that will determine the LATEST FINISH time of the activity in question.

Provided it is done correctly it will soon be clear that some activities have the same earliest and latest start times, and also the same earliest and latest finish times.

These activities are the ones which cannot be delayed without extending the overall completion time of the project. They are, therefore, 'critical' activities and all are linked together to form the 'critical path' through the network, marked by the double line in Figure 6.6b.

The definition of a critical activity is that it is an activity on the critical path which, if delayed, delays overall completion by an equivalent period of time. It is important to note that the definition refers to an extension and not reduction in duration. Although it is possible that a reduction in activity duration may lead to an equivalent reduction in overall project completion time, it may not because reducing the duration of an activity on the critical path can create a new critical path which excludes the activity in question.

Activities not on the critical path will have different earliest and latest start and earliest and latest finish times. The difference in these times is called the float time. The float is the time which the activity can be delayed beyond its EARLIEST FINISH time without delaying overall project completion.

More accurately, such float would be described as the TOTAL FLOAT time and is determined by applying the formula:

$$\begin{matrix} \text{EARLIEST} \\ \text{START} \\ \text{TIME} \end{matrix} - \begin{matrix} \text{LATEST} \\ \text{FINISH} \\ \text{TIME} \end{matrix} - \begin{matrix} \text{ACTIVITY} \\ \text{DURATION} \end{matrix} = \text{TOTAL FLOAT}$$

The total float time can be calculated for each activity in this way, but it is important to note that only one activity in each section of the network can be delayed by its total float. To do otherwise would result in a delay in project completion time. Total float time calculations are given in Table 6.3a.

Table 6.3 Calulation of float times

(a) Total float

Activity	Latest finish time	−	Earliest start time	−	Duration	=	Float
a	4	−	0	−	4	=	0
b	10	−	4	−	6	=	0
c	13	−	10	−	3	=	0
d	15	−	10	−	2	=	3
e	25	−	13	−	12	=	0
f	25	−	13	−	7	=	5
g	25	−	13	−	4	=	8
h	28	−	25	−	3	=	0
i	19	−	12	−	4	=	3
j	28	−	16	−	9	=	3
k	34	−	28	−	6	=	0

(b) Free float

Activity	Earliest start time of succeeding activity	−	Earliest finish Time of Activity	=	Float
a	4	−	4	=	0
b	10	−	10	=	0
c	13	−	13	=	0
d	12	−	12	=	0
e	25	−	25	=	0
f	25	−	20	=	5
g	25	−	17	=	8
h	28	−	28	=	0
i	16	−	16	=	0
j	28	−	25	=	3

There is another float time which is of value for estimating the FREE FLOAT (see Table 6.3b). This float gives a better picture of the amount of free time available in the network as a whole. It indicates the amount of free time available in any one section of the network and, as can be seen, it is applicable only to activities at the end of a network path. Free float is calculated by the following formula:

FREE FLOAT = EARLIEST START TIME of succeeding activity minus EARLIEST FINISH TIME of activity

Floats are particularly useful in resource scheduling, as discussed later in this chapter.

Interpreting the network

The completed project implementation network provides management with significant information about the project. It allows the manager to identify those project activities which will require most careful attention during implementation, for example those on the critical path. This may involve ensuring that the activity resource requirements are given highest priority and that the most able managers are allocated the responsibility for their implementation.

This is not to suggest that other activities should be in any way neglected, but clearly the critical activities will require most managerial attention. This may seem obvious, yet there are numerous examples of project management failing to undertake even a basic networking exercise and allocating priority to activities that are not critical.

For example many rural development projects involve the construction of offices, housing and service facilities. Management is often particularly concerned with setting up the office facilities so that the project can begin functioning and look as if it is in business. Quite often, though, the office facilities remain empty and under-utilized because there is no staff to work in them. Many professional staff are not prepared to move to jobs in the rural areas in the absence of reasonable housing. Staff recruitment and the arrangement for housing staff are activities that are often critical for such rural-based projects, whereas office building is not.

It could be argued that a problem such as the above could be easily resolved without the need for networking. This may be so, but in a complex situation the sequencing schedule of activities is not obvious and the purpose of the technique is to provide managers with information which is based on more than hunches and guesswork.

Networking also has an influence on project design. Once the network is complete it may be that, based on the best estimates of activity duration, the project cannot be completed in the planned time. This may be critical to the feasibility of the project and may raise the issue of whether the project should go ahead in its currently planned form. The result of the network analysis may necessitate a re-think as to the means of achieving project objectives and this could in turn result in a revised project design.

It is rare for there to be only one method of doing a job, but different ways will require different mixes of resources. If time is the critical resource for a particular activity, then this can usually be substituted for other resources at a cost. This process is discussed more extensively later in this chapter, but the issue might arise in the following way with respect to the sugar plantation project.

In this example one of the critical activities is the drilling of boreholes to provide smallholders with clean water supplies. It would not be feasible to move people on to the plantation until the domestic water supply system is complete. However, it might be possible to complete this work more quickly by allocating extra resources to the task. Much will depend on how important it is to complete the whole project on time. If, for instance, there was a need to complete the settlement of the scheme a few weeks sooner in order that the smallholders could plant their crops before the growing season, then clearly a good case could be made for speeding up project implementation. The cost of not completing the project on time may be a year's lost sugar production – a high cost to the project and smallholders alike.

Further implementation planning procedures

The completion of the initial network analysis leads to a series of other tasks which can be undertaken in preparation for project implementation. For illustrative purposes these are dealt with in the following order:

1. bar charting
2. activity description sheets
3. resource smoothing
4. resource budgeting.

Bar charting

Once the logical sequencing of a set of project activities has been drawn up and the networking exercise completed, it is possible to depict the implementation plan in the form of a chart. This is done by simply listing the activities along the vertical axis and indicating their place in the schedule against the time periods along the horizontal axis. Generally the activities are listed downwards in order of precedence, starting with the first activity at the top left-hand corner of the graph. See Figure 6.7 which is derived from the land development component of the sugar plantation project.

The activity start times on a bar chart are usually earliest start times determined from the network analysis. The length of the solid bar depicts the expected duration of the activities and the dotted line following some of the activities indicates the total float for each of the activities not on the critical path. Activities on the critical path have, of course, no float.

The bar chart, or Gantt chart as it sometimes known, is a very useful visual aid to project implementation. The activities which will be taking place at a given point in time can be seen at a glance and, therefore, the chart can be used as a basis for project monitoring. As implementation

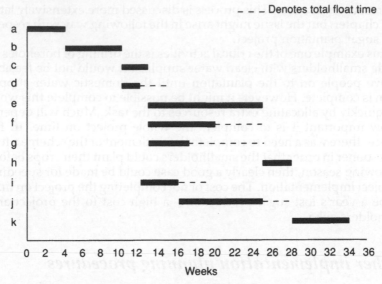

Fig. 6.7 Land development component: bar chart

proceeds it will be possible to determine whether the project is on schedule or not by indicating milestones on the chart which depict the beginning or end of important activities. If the milestones are not reached by the planned time, then there will be a need to determine why and take remedial managerial action. (Project monitoring is discussed more fully in Chapter 9.)

However, in order to take decisive action managers need to have a sound knowledge of what is involved in each major activity. This includes knowing who was responsible for implementing the activity and what resources are employed. This information can usefully be contained in a set of activity description sheets.

Activity description sheets

By definition each project activity will be of a nature that can be identified as discrete in terms of its resource use and the type of work involved. It should be possible to determine what these resources are for each activity, both materially and also in financial terms. Most project activities involve the use of a combination of resources such as labour (skilled and unskilled), material inputs, machinery and equipment. These items can be quantified and costed on a day-by-day, week-by-week or month-by-month basis. The activity description sheets are simply a set of documents containing all this information in a systematic manner. An illustrative example project from the case study is given in Table 6.4.

Table 6.4 Activity description sheet: example

Activity No.	f	Land development work	
	Project		Project
Start Date	week	Finish date	week
Scheduled	13	Scheduled	20
Actual		Actual	

Responsibility for: Post

Authorization -   Project Manager
Supervision -   Deputy Project Manager
Implementation -   Mobilization Team Leader

Resource requirements

1 Labour week	Units per week	Rate/Unit	Total/Week	Weeks Duration	Total Activity
Skilled	2	50	100	7	700
Unskilled	8	30	240	7	1680
Sub total			340		2380

2 Materials					
Aggregate	5 Tons	20	100	"	700
Cement	2 Tons	75	150	"	1050
Fuel	250 Litres	2.50	625	"	4375
Sand	20 Litres	35	700	"	4900
Sub Total			1575	"	11025

3 Equipment					
Pick up	1	120	120	"	840
Trucks	2	400	800	"	5600
Mixer	3	60	180	"	1260
Bulldozer	1	500	500	"	3500
Sub total			1600	"	11200

| Activity total | | | 3515 | | 24605 |

Activity description sheets have a number of uses. First, they can be used to assist line managers plan their work within the resource constraints indicated on the sheet. If these are too restrictive then it will be up to the managers to argue for more resources at the planning stage and this will assist in the overall budgeting exercise.

Second, the sheets can be used to assist project personnel in determining what their respective responsibilities are for each project activity. In the example three levels of responsibility are given: authorization, supervision and implementation. While it is unlikely in reality for a manager to be unaware of who is responsible for each project activity, it may help to improve communications within the project organization if these responsibilities are set down on paper. This will avoid the problem of project personnel claiming that they were not aware of their responsibility towards a particular activity in the event of something going wrong with implementation. Clearly it would be desirable for the indications of responsibilities in the activity description sheets to match those in the linear responsibility chart discussed in Chapter 4.

Finally, the activity description sheets can be used as a basis for altering the activity schedule in order to reduce the overall cost of the project or eliminate bottlenecks in resource use and speed up overall project implementation time.

Resource smoothing

When a number of activities are taking place simultaneously there will inevitably be a high demand placed on project resources. This may be crucial when several activities require the use of a particular project resource which is either very expensive or in limited supply. Transport is perhaps the most constraining resource in many development projects and can serve as a useful example to demonstrate the resource smoothing technique.

In the example project it can be seen in Figure 6.8a that if all the activities begin on time at their earliest start dates, three activities during project weeks 14 to 17 demand the use of trucks. Drilling boreholes and waterway construction each require a single truck to service the requirements of the activity and land development work requires two. Thus during the period week 14 to 17 four trucks would be required to keep all three activities going.

If there was no constraint on the availability of trucks, either owned by the project or freely available to hire, then this scheduling of activities presents no problem. However, if there was a constraint, as might be expected, then it would be reasonable for management to try and determine whether or not rescheduling project activities might be possible in order to avoid purchasing or hiring additional trucks at considerable expense.

In the simple example of the sugar project an examination of the bar chart in Figure 6.7 indicates that activity 'g', waterway construction, has a long free float time, some eight weeks in fact. It would, therefore, be quite feasible to delay the start of this activity by several weeks so that it does not take place at the same time as the land development work. Postponing the waterway construction by seven weeks would remove the need to have four trucks working on the project at any one time.

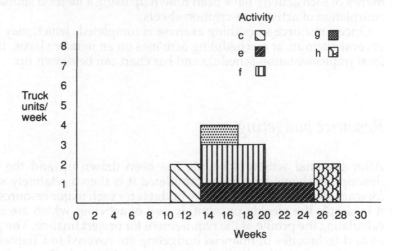

(a) Resource use histogram for truck hire earliest start tine of activities

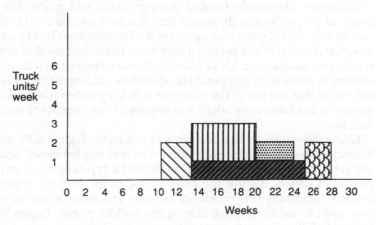

(b) Resource use histogram with activity delayed by 7 weeks

Fig. 6.8 Resource scheduling and smoothing

It is important to note that not only has this simple rescheduling exercise removed the requirement for a fourth truck, with a consequent saving of costs, but also the project can still be completed on schedule provided all the other activities are completed on time. Delaying the construction of the waterways has not caused any delay in the start of its succeeding activity.

103

This resource smoothing exercise can be done systematically only if the network analysis has been done correctly and if the resource requirements of each activity have been drawn up using a method similar to the completion of activity description sheets.

Once a resource smoothing exercise is completed, which may involve several attempts at rescheduling activities on an iterative basis, then the final implementation schedule and bar chart can be drawn up.

Resource budgeting

After the final activity schedule has been drawn up and the activity description sheets have been completed it is then a relatively straightforward step to draw up resource budgets for each major resource. Often it is simply the weekly or monthly cost of activities which are used for calculating the profile of the requirement for project finance. The purposes and techniques of financial budgeting are covered in Chapter 8. The purpose of including the budgeting exercise here is to demonstrate that project budgets should be drawn up on the basis of a realistic plan arrived at through a systematic approach.

Availability of project finance is invariably a problem. Even where projects are adequately funded through loans and grants the management of project funds demands that the best use be made of them. It would be a sign of poor management if all project loan funds were drawn down at the start of the project if they were not in fact needed until several weeks or months later. Once loan funds have been drawn down they are subject to interest charges and are, therefore, adding to project costs. It is axiomatic that the job of the manager is to keep down costs. Making sure money is available only when it is required is an important management function.

Using the sugar plantation project example, Figure 6.9a depicts the financial requirement of each activity as it is implemented according to the schedule in the bar chart. This depicts a typical project expenditure profile in which, during the early phase of the project, expenditure is limited. Gradually this builds up as more and more activities get underway until it reaches a peak during the middle phase. Expenditure then begins to fall away as the project nears completion.

Figure 6.9b shows the cumulative expenditure profile over the duration of the project. Such profiles are generally called 'S' curves. The left-hand vertical axis shows the actual total expenditure in currency terms. The right-hand vertical axis shows the percentage of total project expenditure which has occurred at any point in time. The project time periods are depicted on the horizontal axis.

The drawing up in this way of the project financial requirement serves two main purposes. The first is a planning function in that it provides management with information as to how much money is required to service the needs of the project and when it will be needed.

The second purpose is that once the project has begun the expenditure profile drawn up at the planning stage can be used as a bench-mark for monitoring project implementation. Measuring variances from the planned expenditure profile is a major tool of project monitoring, supervision and control as discussed in Chapter 9.

(a) Financial budget based on activity
    description sheets

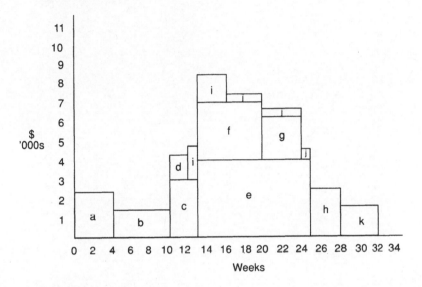

(b) Cumulative financial requirement

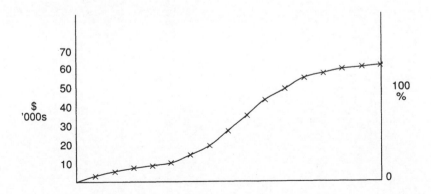

Fig. 6.9 Financial budget and cumulative requirement

References and further reading

Austen A D, R H Neale 1984 *Managing construction projects: a guide to process and procedures.* Geneva, ILO.

Lock D 1984 *Project management.* Aldershot, Gower.

Mulvaney J 1978 *Analysis bar charting.* Basingstoke, Mulvaney.

Smith P 1984 *Agricultural project management: monitoring and control of implementation.* London, Elsevier.

CHAPTER 7

Procurement, contracting and the use of professional services

Introduction

Procurement is a major part of the process of project implementation. It determines who will actually perform much of the work required by the project, and under what conditions. It is an important element of the project manager's responsibilities. In the past, the term 'procurement' was applied specifically to the purchase of goods and equipment, but its use has now been widened to the obtaining of goods, works or services from others in pursuit of the project objectives. Goods may mean anything from the purchase of nuts and bolts to the supply of a complete capital plant such as a processing facility; works covers the whole field of civil or infrastructure development such as buildings, roads and utilities. Services most commonly means the provision of consulting services but could also cover, for instance, accountancy or inspection services. Procurement strategy and practice influences the nature of the relationship between the project and its environment by determining the different responsibilities of the various parties involved and the extent of the project owner's direct dealings in the environment (for example, in arranging the supply of materials or in being responsible for acquiring and utilizing new technology). It is worth emphasizing at this point that there are two basic options available for obtaining goods and works, and undertaking services. Either this is done in-house by the agency who will eventually own the project, or it is done by another party which enters into some form of contract with the owner to perform the work. The first alternative, variously known as 'force account' or 'direct labour', is commonly used for activities which cannot be measured in advance or which must be co-ordinated with ongoing activities (such as maintenance), for work requiring special expertise or which has social objectives such as reducing unemployment, or where an agency is attempting to develop in-house capability. However, there are dangers in in-house work, including, among others, vague definition of work requirements, political interference, overstaffing and lack of financial discipline; at the present time there is a trend away from carrying out activities in this way if it is possible to obtain them from others under contract. This chapter examines the process of procurement under contract.

Each type of procurement has many particular and specialist features but certain common principles underlie them. Throughout this chapter reference is made, on the one hand, to the project 'owner', for whom the project manager is acting, and on the other hand, to the 'contractor' as the party who is supplying the goods, building the works or performing the services.

Procurement strategy and planning

Objectives and options in procurement

The primary objective of all procurement, public or private, is to obtain the goods or work in the optimum and most economical fashion, taking into account the requirements of quality, price and programme. Owners will also be concerned to maximize the probabilities that the contractor will not go out of business or otherwise drop out of the contract, that there will be no costly disputes, and that after-sales service, if needed, will be available on satisfactory terms. It will also be borne in mind that this particular project may not be the only transaction in the owner–contractor relationship. Over and above these, public procurement agencies may have additional objectives. Because they are in the public domain, they are concerned to assure accountability to the sponsoring ministry, the government and society. They may also be required to further national policy goals (for instance, the promotion of domestic contracting capability). Finally, public procurement agencies are often concerned to comply with aid agency requirements when undertaking aid-funded projects. Fulfilment of these multiple objectives means that public procurement is a complex and often lengthy process.

For work procured under contract, three broad conditions can be identified. The first may be termed the buyer's market, where there is significant competition to undertake the work and where the buyer (owner) can determine the general conditions under which the activity will take place (these will of course be modified somewhat during contract negotiations). The most well-known method of procurement for development projects in these conditions is international competitive bidding (ICB). This method is favoured by most multilateral agencies and is designed to make the process of winning the contract open and fair. Features of ICB are notification of bid to all potential bidders, use of neutral design standards and specifications, and precise criteria for the selection of the successful bidder. ICB is favoured because competition should promote economy and because it should be easily accountable, thus preventing corruption and favouritism.

Local competitive bidding (LCB) is a variation of ICB using for small contracts, particularly those involving labour-intensive goods or works, and those scattered geographically or over time. Such contracts are not likely to attract international interest and competition. LCB may also be used specifically to promote local industry, particularly when aid funds are not required to finance the procurement.

Bidding, however, implies a competitive situation and may not always be applicable: competition may not be appropriate for minor orders, repeat orders, extension of works, or in conditions of great urgency or a need for security. Competition may also be unavailable, in the case of proprietary works or goods with advanced technological components.

A second situation may therefore occur if, for any reason, competition is not appropriate but both owner and contractor have a balanced interest in the activity; for instance, an extension of works or standardization on a particular manufacturer's equipment to improve operational efficiency. In this case negotiated procurement will take place, leading to a negotiated contract. It is worth noting that this is becoming quite common for very large contracts, when an element of project financing is often included. For instance, major infrastructural facilities are sometimes acquired under the 'Build, operate and transfer' (BOT) system in which the contractor arranges the finance, designs and builds the facilities, and then operates them for a limited period, after which they are transferred back to the sponsoring agency. The returns to the contractor's investment are obtained through collection of appropriate revenues or other similar arrangement.

Finally, the situation may occur when the buyer is operating in a seller's market and the seller (contractor) will determine the conditions under which the contract takes place. At its simplest level this is the case in 'shopping contracts', particularly used for small items of equipment. However, it may also be used for much larger projects, especially if proprietary items are involved or if, for any reason, a vigorous contracting market does not exist. In these conditions, owners may sometimes benefit by inviting offers, including contract terms drafted by potential contractors. Comparison of these helps owners to learn in greater detail what issues are involved before entering into a final round of talks in which they seek to select the best aspects of the offers made and persuade all potential contractors to conform to them, prior to a bidding process.

Contract planning

Several factors must be taken into account in deciding what procurement conditions pertain, in making the basic choice between in-house work and contracted work, and in packaging the various elements of the project into individual contracts or purchase orders. Chief among these are time, functional acceptability, institutional capability, market structures and commercial practices, size, use of local resources, flexibility, and risks.

Time

One aspect is the time required for completion of procurement. If rapid implementation is required, it may be better to hand the entire responsibility to a single contractor, or to do the work in-house. (The procurement process for a large number of individual contracts can be very time-consuming.) On the other hand, goods and works required at different times should be packaged separately because this spreads the administrative burden, allows flexibility (the owner can modify requirements later) and prevents delays building up and affecting the whole programme.

Functional acceptability

The functional acceptability of the services, works or goods being procured is of prime importance. Owners and their managers will need to take a view as to whether their staff are capable of specifying and ensuring acceptability, or whether the services of others are necessary to achieve this. In particular circumstances owners might wish to make use of the contractors' own expertise, in which case some form of turnkey contract would be appropriate (discussed later in this chapter). Functional interdependence is also important; for instance machinery which works in sets such as pumps and engines may more conveniently be obtained from the same contractor, who is thereby made responsible for ensuring that they work satisfactorily together.

Institutional capability

The degree of owner involvement required will be one of the factors determining whether in-house capability is used, or whether single or multiple contracts are let. Owners may wish to involve their own staff as much as possible in the management of the implementation of the project in which case they would aim to be fully involved throughout the procurement process and might well work through a large number of individual contracts. In this situation, it may be necessary for the implementation stage to include a large training element so that the owner's organization will be better equipped to undertake similar jobs on its own account in the future. This will affect the division of responsibility and obligations between the owner (including the owner's manager or advisers), and the contractor or supplier. Conversely owners may seek to minimize their own involvement by contracting with others to manage the procurement on their behalf or indeed letting all the work out to a single contractor.

Market structures and commercial practices

Generally a procurement package should include items only from one source of supply. This is because it is likely to be more expensive and less efficient to deal with a supplier who is buying from others.

In certain circumstances, however, the converse may be true and there may be advantages to using a single contract, for instance, in the situation where the contractor has proven specialist expertise in general procurement, or the owner's institutional capability is not sufficient to handle a large number of different contracts.

Commercial practices in regard to transportation, assembly and erection must also be taken into account in determining how to split project items into individual packages.

Size

The size of the package must be carefully selected bearing in mind the international and domestic contractors whom the owner desires to interest in the contract. Generally, the bigger the contract, the greater will be the international competition. Conversely it may be desirable to break the project items into smaller packages (this is called 'slicing and packaging'), or to separate out those items which can be manufactured locally, in order to encourage domestic industry. In the case of equipment procurement, the same items (say vehicles) required for different components of the project may be packaged for standardization, convenience and the possibility of discounted prices.

Use of local resources

Procurement may be planned in such a way as to ensure the maximum use of local resources, both natural and human. Slicing and packaging, described in the previous section, is one example of this. Another would be the requirement that a local supply of materials, such as cement, is used during the construction stage; this could be achieved through a nominated subcontract for supply. Yet another example would be the promotion of the skills and capabilities of domestic contractors through the application of the concept of domestic preference, so that foreign expertise will not be required in the future. This applies to the use of labour-intensive methods in countries where labour is abundant (and capital is often conversely scarce). Projects such as road construction can sometimes be broken down into small contracts which are suitable for labour-only contractors (another example of 'slicing and packaging').

Flexibility

Once contracts are entered into they become relatively inflexible, and allow the owner little opportunity to change scope to suit changing conditions. If an owner desires a great deal of flexibility, therefore, it may be necessary to consider undertaking the work in-house. Alternatively, if contracts are to be used, then a number of contracts, perhaps covering a small part of the project over a short time-span, will be more appropriate than one single contract, since the owner will be able to be more flexible between individual work packages. In some situations it may be necessary for procurement, or even construction, to commence before design is complete. In this case, the procurement method must be flexible enough to allow for the effects of design changes during the construction phase through some kind of 'design and build' arrangement.

Risk

Finally, there is the important consideration of risk. There are many political, financial and technical risks inherent in project development, which may cause difficulties, delays and cost over-runs. Procurement planning and practices will determine how these risks are shared between the owner and the contractor, and the relationship between the owner and the environment, not only through the packages into which the project work is broken down, but more particularly in the type of contract that is let. Owners may be concerned to minimize the risks to themselves and place them all with contractors. (Contractors, of course, will take a view on the likelihood of the risky event occurring, and adjust their prices accordingly.) On the other hand owners may be concerned to share some of the risks with contractors, in order to minimize overall costs.

Types of contracts

Different types of contracts are available to take into account the various factors which affect procurement planning, particularly those relating to owners' involvement and risk (see Figure 7.1). First, it is possible to distinguish between those contracts evaluated according to price and those contracts evaluated according to cost.

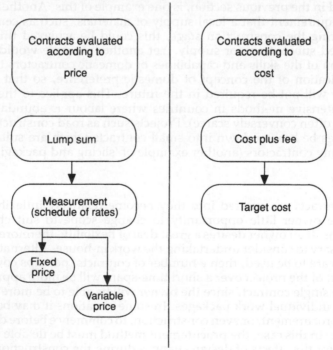

Fig. 7.1 Contract classifications

Price-based contracts

Price-based contracts are appropriate to those conditions where competition exists and the owners' requirements are known in sufficient detail to determine the overall scope of work before the contract is let. Price-based contracts can be lump sum, in which all the risk is borne by the contractor. Otherwise the more familiar 'measurement' contract will be used, in which the actual quantity of work done is measured as the basis of payment to the contractor. Both lump sum and measurement contracts can be on the basis of fixed price, or variable price. In the latter case, where the price is adjusted in line with previously agreed price indicators, the risk of price changes is shared with owners, rather than borne solely by contractors.

Price is the most common basis for awarding and administering contracts. It has the advantage of being simple and promoting competition (provided that a vigorous contracting market exists). On the other hand, price-based contracts generally mean that the owners' and the contractors' financial objectives are in conflict; owners want the maximum possible, while contractors want to provide the minimum permissible, for the agreed price. Contractors, in addition, will often add a large amount for risk contingencies while emphasis on price may conceal other important aspects of undertaking the work, thereby decreasing owners' ability to manage the project in their own interests.

Cost-based contracts

Circumstances favouring cost-based contracts include a low level of work definition at the time of tender (which may be brought about by a need for rapid implementation), work of exceptional technical or organizational complexity, and works involving major unquantifiable risks. In this case the contractor is paid against costs incurred, plus an allowance for overheads and profit. The problem with this type of 'cost plus fee' contract is that it contains no incentive for contractors to minimize costs. Therefore variations of this type of contract have been tried where a 'target-cost' is negotiated between owners and contractors. The contractors' fee is increased when the final cost is below the target cost. A target-cost contract was used for the construction of the Songea–Makambako road in Tanzania and judged to be quite successful. The main problems with 'target cost' contracts are, first, how to determine the target cost, and second, how to prevent problems arising from delays, which assume a much greater importance than in other types of contracts.

While it is valuable to distinguish between price-based and cost-based contracts, it should not be thought that contracts have to be exclusively of one sort or the other. Although, generally speaking, the value content is predominantly either one or the other, contracts can contain elements of the various types, for instance lump sum, measurement and cost plus.

A further distinction which can be made between types of contracts is between those in which design and performance of the contract is by different parties, and those in which design and performance is by the same party. The former category comprises the traditional works and services contracts, in which the owners (or someone acting on their behalf) design and specify the scope of work in sufficient detail for the potential contractors to bid and then undertake the contract, under the intensive inspection and administration of the owners or their managers.

If owners do not want to be involved at the design stage, or do not have sufficient expertise, then some form of turnkey contract will be appropriate. A whole range of these contract types can be distinguished, from the situation where contractors design and construct the complete operating asset, and owners turn the key to start it, to situations where owners have a much greater involvement through the design period, and contractors are only responsible for providing part of the design services, perhaps relating to the central process or technology.

Turnkey contracts have the advantage of providing continuity of technical and administrative responsibility, and reducing the number of contractors required. They may also provide an inducement to contractors by joining an attractive contract to an unattractive contract. On the other hand turnkey contracts may have a high cost. They also provide little or no opportunity for owners to become involved in the design and construction stages, thus making it more difficult for them to become familiar with the operating characteristics of the facility.

Each type of contract has different attributes related to the various factors discussed affecting contract planning. Turnkey contracts, for instance, imply little involvement by owners, minimum flexibility and (perhaps) faster implementation. Contracts which allow variation both in the quantity and price of work place part of the contract risks with owners, while lump sum, fixed price contracts place all the risk with contractors. A number of small contracts will allow owners to develop and utilize domestic contractors. Selection of the appropriate procurement strategy and contract type will involve owners and their managers in judgement and trade-off between the various factors.

Managing procurement lead time

If managers judge that a buyer's market exists and that there will be competition among potential contractors to undertake the project work, then a bidding process will be required, leading up to the award of the contract. The overall programme is thus made up of procurement lead time and contract performance. Time for contract performance depends on the nature of the contract and can vary from days to years. Procurement lead time is made up of a number of separate steps surrounding the central bidding process, (Figure 7.2), the most important of which should be the following:

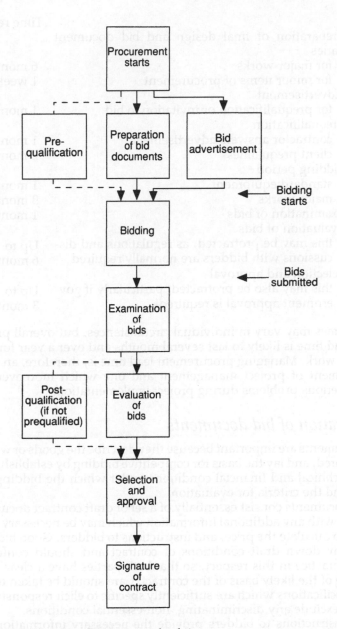

Fig. 7.2 Procurement lead time

		Time required
1.	Preparation of final design and bid document varies	
	for major works	6 months
	for minor items of procurement	1 week
2.	Advertisement	
	for prequalification or invitation to bid	1 month
3.	Prequalification	
	contractor answers advertisement	1 month
	client prequalifies	2 month
4.	Bidding period	
	standard equipment	1 month
	major works	3 months
5.	Examination of bids	1 month
6.	Evaluation of bids	
	this may be protracted, as regulations and discussions with bidders are normally required	Up to 6 months
7.	Selection and approval	
	this may also be protracted, particularly if government approval is required!	Up to 3 months

These times may vary in individual circumstances, but overall procurement lead time is likely to last several months and over a year for major items of work. Managing procurement lead time is, therefore, an important element of project management and one which moreover often causes serious problems during project implementation.

Preparation of bid documents

Bid documents are important because they describe the goods or works to be procured, and lay the basis for competitive bidding by establishing the legal, technical and financial conditions under which the bidding takes place, and the criteria for evaluation.

Bid documents consist essentially of a set of draft contract documents, together with any additional information which may be necessary for the bidder to calculate the price, and instructions to bidders. Good bid documents lay down draft conditions of contract and should conform to market practice in this respect, so that both parties have a clear understanding of the likely basis of the contract. Care should be taken to write clear specifications which are sufficiently specific to elicit responsive bids and yet exclude any discriminating, non-essential conditions.

The instructions to bidders provide the necessary information as to how bids for the work are to be prepared and submitted. Typically, instructions to bidders will provide some background information on the project and the source of funds, the contents of the bid documents, how the bids are to be prepared (language, price, bidder's eligibility, validity), submission procedure, bid opening and evaluation (including domestic preference, if applicable) and award of contract. A model set of instructions to bidders may be found in 'Procurement of Goods', sample bidding documents produced by the World Bank and Inter-American Development Bank (1986).

Advertisement

Advertisements are required to give prospective bidders notice of the prequalification process and of the invitation to bid itself. The objectives of the advertisement are to obtain the widest possible competition and to ensure a fair treatment of all prospective bidders (it is often required by international lending agencies for the latter purpose). The advertisement should give brief details of the goods, works or services required, procedures for purchase and submission of bid documents, and eligibility to bid.

Suitable methods of notification and advertising are local and foreign newspapers, trade and technical journals, official notice boards and gazettes, and through embassies and chambers of commerce. It may be possible to include several bid packages in one advertisement.

Capabilities of bidders

At some point before the contract is signed, it will be necessary for owners to assure themselves that contractors are technically and financially capable of undertaking the contract. While many problems are encountered during the procurement lead time, projects suffer even more serious delays if contract performance is unsatisfactory or has to be terminated. Bidders' capabilities may be examined during prequalification after which only 'prequalified contractors' are eligible to bid. The process should take into account the contractors' experience in similar work in developing countries, capacity with respect to personnel and equipment (if appropriate), financial status and existing commitments.

Prequalification, though sometimes a lengthy process, has several distinct advantages. Its objective is to ensure that only reputable contractors are allowed to bid and thus

1. it saves unqualified bidders the time and expense of preparing bids which would subsequently be rejected
2. it ensures that qualified bidders will not be deterred from bidding by the fact that they will be unrealistically undercut by unqualified bidders
3. it reveals the amount of interest in the contract.

When evaluating responses to a prequalification invitation it is best to use a merit point system, agreed in advance, for each feature. The objective should be to prequalify all firms achieving more than an agreed minimum of points (including possibly a minimum score for each criterion) and not to carry out any sort of ranking exercise. In most internationally funded procurements, prequalification is not down to a pre-set number. All who are capable should be allowed to bid.

117

Prequalification need not normally be used in supply contracts. In that case it is possible to use post-qualification to check on the bidder's financial and business position and therefore to save time by looking at only two or three sets of documents. For works and turnkey contracts on the other hand, prequalification performs a useful function.

It may not be necessary to carry out prequalification for every new contract. Instead a short list of prequalified contractors can be maintained. If prequalification is not used, then the most acceptable bidder must be post qualified before the contract is made.

The Bidding Period

The bidding period is one of intense activity for the bidders as they acquire the necessary information on which to put together their bid. Less activity is required from owners or owners' project managers during this time. However, it may well happen that individual bidders will come to owners seeking clarification on points in the bid documents: in keeping with the principle of fair and equal treatment to all bidders, owners must ensure that requests for clarification, and owners' responses, are communicated to all who purchased the bid documents.

Bid opening

At the end of the allowed period, bids are submitted at the specified place and time. The owner should return, unopened, all bids not submitted in accordance with this specification. There then follows the bid opening, again at an officially notified time and place.

With regard to bid opening procedures it should be borne in mind that, although public bid opening is the generally accepted method since it can be seen to be fair, there are sometimes advantages in secret bid opening. Secret bid opening prevents undue pressure being put on officials involved in the evaluation and it may also help to prevent collusive bidding by a cartel, because the identity and prices of the bidders are not known by other bidders at the time of bid evaluation. Another variation of the bid opening procedure is to use the two envelope system (commonly used for evaluating proposals for consulting services) in which the technical bids in the first envelope are examined and evaluated first. The second envelope containing the financial prices are then opened for those bids which are found to be technically sound.

Bid examination

Preliminary examination of bids is necessary to ensure that they are complete, valid, substantial, responsive and, if accepted, would lead to a contract. This is normally a fairly rapid process carried out by project managers who are required to check such matters as the correctness of the bids submission, bidder's eligibility, mathematical computations and the like.

Bid evaluation

Bids found satisfactory in the bid examination are then passed for bid evaluation. The objective of bid evaluation is to select the technically responsive offer which is to be considered to be the most economic when assessed over a reasonable period, provided that the capital costs of this offer are such that they can currently be afforded.

In this activity price is only one criterion to be considered, particularly in the case of goods and equipment. The purchase price for plant and machinery is only a small part of the total ownership cost, typically 25–50 per cent. It is therefore necessary to try to make some comparable estimates of total ownership costs in order to determine the most advantageous bid. Other factors which may therefore be important are the delivery or completion schedule, terms of payment, operation and maintenance costs, efficiency, reliability, resale value and guarantees. The criteria to be used for evaluation must be clearly stated in the bidding documents in order to allow the bidders to respond and so that they provide the necessary information, for instance relating to the elements of domestic manufacture. If possible the method of allowing for the various criteria should also be stated so that the evaluation process can be seen to be fair. Evaluation of bids can be a complex process, particularly for sophisticated equipment or works, and the reader is referred to specialist texts (for example, Westring 1985) for their treatment.

Contracts

Contract documents

Contracts used in project implementation are almost invariably written contracts, the contract being contained in one or more volumes of contract documents. If the contract has been arrived at through a bidding process, the contract documents would have been available in draft form at the time of bidding and will have been modified and agreed during negotiations between owners and the preferred bidder. Contract documents comprise some or all of the following major items:

1. technical specification
2. bill of quantities
3. price schedules of plant
4. conditions of contract
5. contract forms.

The specification contains a description and design information for the works, drawings including layouts and site plans, performance standards to be achieved, plant operation and maintenance procedure, training requirements, inspection and tests on acceptance of the works. Specifications for construction tend to be tighter than industrial or plant specifications, which are functional and rely on the know-how of the contractor. The specification, therefore, defines the quality of the goods or work to be provided, in an effort to ensure that it is fit for the purpose for which it is intended by the owner. At the same time, those drafting specifications must guard against over-specification, since the owner must be in a position to afford the initial purchase and subsequent operation of the items.

Standardization is a particular problem in writing specifications. The owner can give precise information in order to standardize on equipment by referring to national or international standards. There are advantages of standardization, as this may help in getting good quality equipment of adequate reliability, but too much stress on standardization may lead to elimination of competition or lack of progress. Moreover, standard specifications (and codes of practice) reflect the lowest generally acceptable standard and are not necessarily adequate for specific applications. There is, however, a case for standard conditions of contract (see below).

Bills of quantities are used in works contracts and are designed to allow the owner to state the estimated quantity of a particular item which is required, while the contractor states the price. Thus an approximate estimate of the contract price can be obtained. (In measurement contract the actual quantity of work done is remeasured when it has been completed and the contractor is paid at the agreed rate for this quantity.)

In a contract for the supply of goods, owners provide a schedule of requirements, which is then priced by the bidders.

There are a large number of model forms of conditions of contract, covering a wide variety of contractual situations (see Figure 7.3). Use of these model forms saves time in the preparation of bid documents. In addition they tend to be preferred by bidders because the risks, obligations and incentives inherent in a particular set of conditions are already known and experienced; in fact for many of them a body of case law has been developed which facilitates contract administration and supervision. Contractual law is complicated by different legal systems, for example the Continental European (codified) law as opposed to the British (evolved) system and this is reflected in different model forms of contract. It should also be borne in mind that model conditions apply only to the general conditions of contract, key clauses of which are discussed on pp. 122–3. Most contracts will also require special conditions, for instance concerning contract programme or delivery period.

In addition to conditions of contract there may be site regulations, particularly for construction contracts. These may govern matters relating to methods of working such as night working, labour laws, safety, etc. As they do not form part of the conditions of contract they do not have the same importance within the contract, though they may have an important bearing on the manner of execution, particularly if they are part of national law.

Finally, the contract documents will contain forms such as the offer and acceptance, and very often a form of agreement as well. The offer and

---

Conditions of contract: model forms

Contracts for the supply of goods

United Nations (1980) *Convention of contracts for the international sale of goods*

World Bank and Inter-American Development Bank (1986) *Sample bidding documents: procurement of goods*

Contracts for the supply and installation of goods

FIDIC (1978) *Conditions of contract for electrical and mechanical works including erection on site, third edition* (the 'yellow' book)

Contracts for works

FIDIC (1989) *Conditions of contract (international) for works of civil engineering construction, fourth edition* (the 'red' book)

Contracts for services

FIDIC (1980) *International model form of agreement between client and consulting engineer for project management* (other model forms refer to pre-investment studies, supervision of construction, operation maintenance and training, etc)

The European Development Fund has also produced general conditions of contract for supply of goods, works and services, reflecting the different legal systems which operate in the European, as opposed to the Anglo-American contracting tradition

Turnkey contracts

UNIDO (various dates) has produced draft model forms of agreements for the construction of a fertilizer plant, under different contract types (turnkey lump sum, cost-reimbursable, provision of know-how and engineering services, etc)

Fig. 7.3 Conditions of contract: model forms

acceptance record the mutual consent of the two parties to enter into the contract and must of course be properly signed and witnessed. The offer will state the bid price, and give additional technical information such as alternative offers, non-conformity with the specification, etc. There may be an appendix to the offer where important details pertaining to the contract such as the period for commencement and maintenance, mobilization payments, and so on, can be recorded.

Conditions of contract

The conditions of contract define the terms under which the contract is performed. These of course will vary in scope and complexity between different types of contract and indeed for each individual contract. For instance, the World Bank model conditions for the supply of goods has thirty-two main clauses, covering such items as the obligations of each party to the contract (purchaser and supplier), delivery, payment, security of performance, changes to the contract, and the like. Contracts for works are, of course, inherently much more complicated to draw up and administer because of the uncertainty and risks inherent in the working environment (particularly related to ground and climatic conditions). The International Federation of Consulting Engineers (FIDIC) model conditions of contract for works, therefore, has some seventy clauses, covering in addition aspects of labour, materials, workmanship, programme and the mechanisms for dealing with unforeseen conditions.

Readers who need to become deeply involved in these matters are referred to specialist texts (Westring 1985). However, key matters that are likely to arise in most contracts are price, payment, delivery and security, and these are discussed briefly below.

Methods of calculating the contract price and making payment are naturally always of major concern in the drafting and agreement of contracts. Contracts may be fixed or variable price. If they are short term (less than one year), fixed price contracts are appropriate. Otherwise a variable-price contract may well result in a lower cost to owners, particularly at a time of high inflation, as it avoids the necessity to contractors of building a high provision for inflation into the price. Variable-price contracts are best administered on the basis of a price adjustment formula, which requires a breakdown of the work into its constituent parts (eg labour and materials) and payment for these elements against nationally or internationally recognized indices of prices. An example of such a price-adjustment clause is given in the World Bank's Sample Bidding Documents for the Procurement of Goods (Clause 12 of the Special Condition of Contract).

Schedules of payments will vary across different types of contracts. For services and works contracts, payment is likely to be made on a progress basis as the work is undertaken. Advance (or mobilization) payments may also be involved, and retention to ensure full performance. For the supply of goods, on the other hand, payment will usually be made on a staged basis, with the bulk of the price (80–85 per cent) being paid when the goods are ready for shipment.

With regard to delivery, INCOTERMS, published by the International Chamber of Commerce (1983), are a commonly used set of conditions describing the respective obligations of owners (purchasers) and contractors (sellers) for different delivery arrangements.

Letters of credit ('banker's documentary credit') are commonly used to make payments for offshore procurement. This is an arrangement whereby a bank (the issuing bank), acting on the instructions of a customer (the owner) makes payment to a third party (the contractor), against stipulated documents. It is important to note that a bank's obligation to pay a letter of credit does not depend on performance of the underlying contract, but only on the presentation of stipulated documents. Owners, therefore, have to take other action, such as employing an inspection firm, in order to safeguard themselves against failure in contract performance. The International Chamber of Commerce (1975) has produced a guide to the use of letters of credit entitled, *Uniform custom and practice for documentary credits*.

Different types of security are available to owners during the procurement process. The two most important ones are bid bonds and performance guarantees. Bid bonds are often required at the time of bidding, to ensure that bidders do not withdraw their bids, or fail to sign the contract if awarded the bid. These are generally for up to 2.5 per cent of the tender price. Performance bonds or guarantees are required to provide owners with security for the proper performance of the contract. These are normally for 10 per cent of the contract price, though they can be much higher in American practice. In addition owners may like to consider the use of advance payment guarantees (to ensure that contractors do not abscond after receiving mobilization advances), retention bonds (so that contractors can be paid full amounts, while still providing owners with security against satisfactory completion of the contract) and maintenance bonds, in lieu of retention during the maintenance period.

Owners can obtain other security for satisfactory performance both through the contract and in common law. Liquidated damages are an estimate of the actual damages suffered by owners if the contract is not completed on time (a bonus for early completion may be similarly applied). Warranties are generally available protection for owners against faulty design and workmanship, both for goods and works. In addition, works contracts can specify a maintenance period (generally one year), when the contractor must remain on site to make good any defects that come to light.

Contract administration

Contracts for the supply of goods

Rejection of unsuitable goods is very expensive for buyers, as well as sellers, especially in offshore procurement. Therefore it is necessary for buyers to take an active role during the performance of the contract. This means inspection and, where appropriate, testing, at key milestones during the contract. At the very least this will probably be done prior to shipment and again prior to acceptance. For goods bought against standard specifications, inspection will be mainly for quantity, packing, marking, and so on, while for goods bought against design specifications, additional inspections and testing for materials, tolerances, etc, may be required at various stages of manufacture. It will normally be necessary to agree the required procedures with suppliers and write them into the contract.

Inspection may be combined with expediting. Both of these functions can be contracted to inspection companies, who will undertake the work for a fee. Many project organizations in developing countries would find it cost-effective to do this. Although the companies' fees can seem large, the potential savings in smooth contract administration are very great.

For complex plant and equipment buyers and sellers need to agree procedures to be followed at the time of commissioning. This has both contractual and operational objectives. Buyers will want to get involved at this stage, in order to gain familiarity with the equipment. The respective rights, duties and responsibilities of the various parties involved need to be worked out at the time of signing the contract, not afterwards.

Administration of works contracts

Works contracts need more constant administration than contracts for the supply of goods. In the Anglo-American legal tradition (the FIDIC conditions), three parties are involved in the contract – the owner, the contractor and the consultant. The contract is made between the owner and the contractor. The consultant is not an employee of the owner (though the consultant will be under separate agreement with the owner) but acts in an independent and quasi-judicial capacity in administering and adjudicating the contract. Often, the consultant is the person (company) who was responsible for the design of the works.

This approach can be contrasted with the European system (used by the European Development Fund), in which the on-site administration is provided by an employee of the owner. This system will also provide on-the-spot contract administration, but does not guarantee the impartiality of the agent. There is a danger that these doubts will lead to delays (while matters are referred back to the employer for decisions) and perhaps also to contractors inflating their prices to cover themselves against risks of disputes during the operation of the contract.

In either system it is desirable that the parties to the contract develop a reasonable degree of trust in one another's good faith and technical competence. No one gains when contracts suffer delays and disruptions (except maybe lawyers!).

Using professional services and consultants

The availability and benefits of professional services

Professional services have traditionally been used in obtaining technical expertise and advice. This was particularly common in the engineering and architectural fields, and often used in the project preparation, design and implementation phases of projects, for instance for the supervision of construction. Nowadays, however, the concept of professional services and consultancy has broadened considerably, to cover sectors other than civil engineering and building, and to include, besides project management and administrative services, a host of other more specified services such as accountancy, legal, marketing, inspection and the like. In addition, we may also find commercial and contracting firms offering their services for the provision of advice: this is inevitable in a time of rapid advances in technology, as sometimes the expertise and technical know-how is confined only to those firms which are seeking to commercially exploit it. Although those selling professional services have tended to stress their impartiality as one of their major benefits, it is not necessarily detrimental to take advice from those with vested interests, as long as those interests are known.

Besides a widening in the scope of professional services, there has also been a widening of the times when such services are engaged. While traditionally this has been confined to project formulation (feasibility), preparation (design) and implementation (supervision of construction), nowadays it is possible to employ consultants and professional services at every stage of the project cycle from identification through to beneficial operation and beyond to evaluation. Increasingly consultancies are offering project and general management services to clients. (Management services can also be provided by contractors under various forms of 'management contracting'.) Consultants may have an important role to play in the final evaluation stage of a project, where their independence from project agencies may be a particular asset.

The most common disadvantages cited in the use of consultants are their high costs, their lack of local knowledge (which may show itself in inappropriate proposals), their inflexibility, the difficulty of transferring their professional skills and knowledge to owners and the difficult working relationships that often seem to exist between consultant and owner. Some of these disadvantages may be overcome by good practice in their appointment and use.

Steps in appointing and using professional services

Definition of strategy

In defining the strategy, the first requirement is to determine whether professional services are needed at all, or whether it is possible to undertake the task with owners' own resources. If consulting services are required these may be obtained from a variety of sources, ranging from the employment of individuals with personal contracts, through the use of university or other academic groups, to the employment of firms of consultants. Here again there are options, in that some firms act simply as recruitment agencies, searching out and engaging personnel on a short-term basis (often directly on behalf of owners), while others keep a large number of permanent staff and specialize in particular sectors or activities. As a general rule these options become progressively more expensive but should, on the other hand, provide the greatest likelihood that owners will receive satisfactory service, or, if unsatisfactory, will be able to have it rectified quickly and at no extra cost.

Drafting terms of reference

This is a vital part of owners' preparation for a consultancy contract, as many of the problems encountered in working with consultants stem originally from poor terms of reference (TOR). The TOR should contain a statement of the objectives of the professional services (not necessarily the same as the project objectives) and the scope of services required. These should set out what is to be done, but not how it is to be done (that is the consultant's job) and should aim for flexibility without ambiguity. Judgement is needed on the degree of detail to be provided, perhaps by specifying person-month requirements of the type of expertise required. The document containing the TOR should define institutional arrangements, clarifying the relationship between owners and the consultant and describing the responsibilities of owners in the performance of the assignment. Reporting requirements and other aspects of the assignment such as work programmes, phasing and sources of data should also be detailed in the document. It is customary to require consultants to comment on TOR when submitting bids since they may well be in a position to give useful advice to owners on aspects of the proposed project.

Evaluating bids and selecting consultants

Prequalification is sometimes used to ensure that only reputable, competent consultants are invited to submit proposals for assignments. It is in any case recommended that only short lists (say five or six) of consultants are invited to submit bids for particular assignments, or the process of evaluation becomes extremely complex.

Three key aspects should be taken into account in evaluating bids and selecting consultants:

1. the expertise and experience of the consultant in the field of the proposed assignment
2. staff proposed for the assignment
3. the adequacy of the proposed work plan.

The relative weights to be given will vary, depending on the type of assignment. Typically 15 per cent of total marks is given for experience, 25 per cent for the work plan and 60 per cent for the staff. In some countries the team leader is given a very large proportion of total marks, even up to 50 per cent. The relative weights to be given will vary, depending on the type of assignment. Other aspects which may be considered in the selection process include experience in the region or country concerned, and the proposed methods of ensuring transfer of technology and knowledge to the client. The use of joint ventures with local consultants is one method sometimes used to achieve this.

It is generally considered that price should not be the prime factor in selecting consultants. However, having selected the most suitable, owners should ensure that all unnecessary expenses are excluded from the consultancy agreement.

Contract administration

It is difficult to make consultancy assignments work. Unlike contracts for works and goods, where the product can be fairly closely defined, the product of a consultancy contract is generally advice, whose quality is hard to assess. Trust, a good working relationship and good communications between client and consultant are obviously prerequisites to success. Good communications do not necessarily mean an extensive system of written reports. These take time to prepare and to read, time which might be more usefully spent on investigation and discussion of the basic issues. Simple points, such as prompt payment, assistance with administrative arrangements and timely review of reports, do a great deal to ensure the smooth running of consultancy contracts.

References and further reading

Fry G W, C E Thurber 1988 *The international education of the development consultant: communicating with peasants and prices.* Oxford, Pergamon.

International Chamber of Commerce 1975 *Uniform customer and practice for documentary credits* Paris, ICC.

International Chamber of Commerce 1983 *International rates for the interpretation of trade terms.* Paris, ICC.

Westring G 1985 *International procurement: a training manual.* Washington DC, UNITAR.

World Bank 1986 *Guidelines for procurement under World Bank loans and credits.* Washington DC, World Bank.

World Bank and **Inter-American Development Bank** 1986 *Sample bidding documents: procurement of goods.* Washington DC, World Bank.

CHAPTER 8

Project finance and financial management

Introduction

This chapter is concerned with a key factor of development projects –
money and its management. Resources such as people and equipment
are essential requirements for the achievement of project objectives, but
they can be deployed within the project only provided finance enables
their availability. A major task for managers is therefore the management
of this crucial resource that fuels the project organization. It can reasonably be argued that ultimate responsibility for managing the project
finances is one of the key factors defining the role of the project manager.

Few development project managers, however, are appointed on the
basis of their knowledge of financial management. This is particularly the
case in the public sector in which projects are implemented within a
financial environment dominated by the government budgetary system.
The established mechanisms for operating the public sector budget are
primarily designed to support departmental programmes and activities
on a recurrent basis. Development projects on the other hand usually
operate outside these mechanisms and are generally one-off investments.
While projects may be set up to operate as semi-autonomous structures
for the purpose of organization and operation, they are often constrained
by the financial procedures employed by the parent department. This is
partly a result of bureacratic conformity and partly because project sponsors sometimes lack confidence in managers' ability to manage project
funds adequately. This is sometimes reflected in circumstances where
'financial controllers' are allocated to projects supported by external
donor agencies. Often there is a three-way contest between the parent
department, the financing agency and project management over the
control of project funds. Effectively this is a contest over who is actually
managing the project.

In other sectors such as the parastatal and private sectors, this division
is less pronounced, but, as development projects normally involve the
use of grants or loans from agencies outside the parent organization,

project managers can be expected to share some decision making with people outside the immediate project. In development projects it is rare for managers to have complete freedom of action over finance. It is this restriction that underpins managers' requirement for a sound understanding of financial matters. A good deal of their time will be devoted to defending expenditure proposals and ensuring that project implementation is not being affected by financial constraints.

The discussion of the management of finance in this chapter is divided into four related areas. These are budgeting, financial management and accounting, investment appraisal, and financial analysis. Budgeting is of crucial importance for the orderly implementation of a project and relates closely to the processes of implementation planning discussed in Chapter 6. Financial management and accounting involves examination of the methods and purpose of accounting and in particular the organizational requirements for project accounting and management of financial information. Investment appraisal, while not the main interest of project managers, is included to provide a link between the appraisal and implementation elements of the project cycle. The fourth area, financial analysis, includes a discussion of the interpretation of accounts and financial statements. The important subjects of project cost control and the monitoring of project finances are discussed in the next chapter, in the context of overall project management control systems.

Budgeting

Development project managers have several major tasks to perform in respect of managing project funds. These can be summarized as follows:

1. determining the requirement for funds to service the needs of scheduled activities during a defined period
2. controlling the project through the allocation of project funds to different activities
3. making adjustments in the allocation of funds between activities during the planned time period.

The basis for undertaking all these tasks is budgeting. At its simplest budgeting is an exercise in estimating the cost of a set of activities and subsequently determining which will be undertaken within the capacity of the resources available. The first task in any budgeting exercise is therefore that of costing.

In Chapter 6 the result of the implementation planning exercises ended with the drawing up of activity description sheets. These indicated the resources required for the completion of an activity and the time period when it would be undertaken. Activity description sheets can readily provide a basis for the budgeting exercise and greatly assist with that of cost estimation.

Cost estimates will have been prepared at several stages during project development, notably prior to appraisal, but will need to be revised for the purposes of implementation: at this stage it may be necessary to carry out detailed design, which can be accurately costed, as well as adding in other costs such as financial contingencies (including allowances for inflation). The aim must be to produce an estimate accurate to plus or minus 10 per cent.

Cost estimation is a notoriously difficult task in many developing countries in which there are high levels of inflation, a dependency on imported goods and relatively poor communications. There is an understandable temptation to overestimate costs in such circumstances in an attempt to hedge against uncertainty. The problem with overestimating the cost of activities is that this may lead to a slow down in project implementation as fewer activities can be undertaken with the given financial resources. This slowdown itself may lead to higher overall project cost.

Equally if project activity costs are consistently underestimated this may result in activities being halted or slowed down as funds run out during the budget allocation period. A common cause of project delay is the over-expectation of what can be achieved with a given amount of funds.

A budget is normally drawn up to cover a set time period, be it a month, a quarter or a year. An annual budget divided into quarterly allocations is a common method. Some activities may spread over several budget periods. Division of the total estimated cost in these circumstances would need to be done carefully according to the homogeneity of the activity. For example any activity involving the initial purchase of plant or equipment would probably require a greater proportion of the budget allocation in the first time period.

Budgets for individual activities would normally be expected to be prepared on a department or section basis according to the size and complexity of the project organization. These would then be aggregated to produce the main project budget based on the implementation plan which in turn reflects the overall objectives of the project. Before the final budget is confirmed there will inevitably be the need for extensive discussions between the manager and section heads to resolve conflicts and issues that may arise through the budgeting procedure. This process enhances the communications between sections and helps all project section heads to appreciate fully the aims and objectives of each other's sections.

In the context of a project there will inevitably be discrepancies between the planners' forecast of financial requirement and the real requirement facing project management at the start of the project. At this point there is likely to be considerable scope for prioritizing and selecting tasks to be carried out against the availability of funds. Here the transformation occurs between 'planning the work' and 'working the plan'. This concept is analogous to that discussed in Chapter 6 in which the implementation schedule, drawn up on the basis of each activity beginning at its earliest start date, is modified during actual implementation to take into account resource constraints and other factors that affect the planned schedule. As with implementation planning, budgeting is not a once and for all activity, but one that is continually being undertaken during project establishment.

Once the preliminary budgets are agreed by project management they may need to be approved by the project sponsors. This may involve considerable revision according to the involvement and perceptions of the sponsor and the terms and condition of the financial package.

Figure 8.1 depicts a typical budgeting process which might apply generally to a development project. This process may take some time to complete. Therefore if circumstances allow, managers should seek approval to have some initial funds released to the project in order to begin some priority tasks. This avoids some of the problems of project start up.

AGGREGATION OF BUDGETS
↓
DISCUSSION OF BUDGETS
↓
NEGOTIATION OF BUDGETS WITH SPONSOR
↓
BUDGET APPROVAL

Fig. 8.1 Implementation planning: preparation of section budgets

Once the budget is approved and funds are made available then the budget takes on a new role. Normally funds would be chanelled through the project accounting section under the control of the accountant who is responsible for dealing with financial transactions and monitoring expenditure against the various budget heads. The accounts section thus becomes to some extent the eyes and ears of the manager with respect to financial matters and regular reporting on disbursement against planned budgets. The adequacy or otherwise of the budgets becomes one of the main tools through which managers can assess progress towards the achievement of project objectives. This is depicted schematically in Figure 8.2. The measurement of actual expenditure against the budget is a fundamental exercise. Further consideration of ways of doing this are discussed in Chapter 9.

RECORDING OF EXPENDITURE
↓
MEASUREMENT OF VARIATION FROM PLAN
↓
REVISION OF BUDGETS
↓
DEFENCE OF REVISIONS
↓
FEEDBACK TO NEXT BUDGET PERIOD

Fig. 8.2 Budget implementation cycle

Using detailed information with respect to activity or section budgets the manager is able to make adjustments to the main budget. Unforeseen requirements for funds invariably arise, some activities cost more than expected and some less, the need to reallocate funds across subheads through virements is a major task of the manager. The scope of the manager to do this is usually restricted in that approval from the sponsors may be required before such reallocation can take place. It is a key task of managers to be able to defend requests for virements and to understand the impact such requests may have on the progress towards project completion.

Precedence and zero-based budgeting

Budgeting is generally a cyclical exercise that takes place on a periodic basis thoughout the project establishment phase. Given that the activities of project implementation are unlikely to be of a regular or routine nature the exercise is more demanding for project managers than departmental line managers.

Line managers rely on the past budget requirements as a guide to those for the future. This is called precedence budgeting and is generally used where departmental activities are of a routine nature varying little from year to year. This might apply particularly to the public sector where staff costs make up the bulk of the budgetary allocation.

Such a method of budgeting by reference to historical precedence may, however, be the source of sustaining inefficiencies within an organization. It may lead to complacency on behalf of managers and deflect them from examining the nature of departmental activities in a critical manner. A budget defence answer along the lines that 'the minimum requirement is last year's budget plus inflation' epitomizes the precedence budgeting method.

'Zero-based budgeting' involves a completely different approach in which each item of the budget must be defended annually from scratch, regardless of the previous accounting period level of expenditure. All items of expenditure need to be justified in terms of the work objectives. This, in theory, ensures that managers will take time to examine the organization and work practices of their sections or departments and thereby identify existing inefficiencies and opportunities for cost saving. There may be considerable advantages to be derived from this process. However, the process can be very time consuming and, more important-ly, it may also result in frequent and significant organizational change which, in turn, may result in considerable stress to personnel.

Generally, for development project implementation and financing the budgeting exercise owes more to the concept and practice of zero-based budgeting than the precedence method given, that by definition, projects involve little which is of a routine nature.

Budgeting as a motivating strategy

So far the discussion of budgeting has concentrated on the issue of estimated and actual financial disbursement, but budgeting plays just as important a role in the behavioural aspect of management. Budgets set limits and targets which can be measured. The allocation of a budget provides freedom of action to the budget holder within the scope of the task in hand. It therefore provides managers with an incentive and an obligation to attain the set targets within the budget. This would seem to be an ideal way of motivating personnel.

However, such an approach might be criticized on the grounds that the motivating force is a negative one, in that budget holders' prime motivation will be to avoid budgetary failure rather than to meet the wider organizational objectives. Failure to achieve the work objectives, failure to expend the budget on time or failure to keep within the agreed budgetary limits, all might be indicators of managerial inefficiency. In fact there may be very good reasons for each of these failures for which the manager is not responsible, not least in the realism of the initial budget allocation.

Project managers might well, on balance, use the budgeting exercise as a means to motivate their staff but it would be as well for them to be aware that failure to meet budgetary targets may cause considerable stress among those staff, particularly when the failures are due to other reasons than inefficiency.

The advantages of the budget system can, therefore, be summarized as follows.

1. It can be directly linked to the implementation planning exercise and therefore directly reflect the objectives of the project.
2. The organizational structure of the project is defined in terms of zones of financial responsibility under the budget.
3. Each project subsection is involved in determining its own objectives and work programme through the exercise.
4. It stimulates communication within the project and helps to resolve conflicts.
5. It provides a basis for allocating priorities and assists with resource allocation throughout the project structure.
6. It provides the basis for a monitoring and control system which can allow for speedy corrective action if required.
7. Budgeting allows management personnel to take responsibility for planning their work, setting targets and measuring their performance which can all add up to a stimulus to obtaining higher levels of achievement.

These advantages make the budgeting process a powerful tool which is vital for effective project management. Good budgeting procedures ensure good communications within the project, provide a motivating impetus and allow project personnel to identify the contribution they are making towards project completion.

Despite these advantages many project managers find the project budgeting exercise to be the most difficult aspect of the project to manage. This applies particularly to public sector projects in which no significant revenue attributes to the project. When the project manager has no control over the flow of funds into the project the exercise can become very frustrating. If the project is reliant on a central procurement agency, or a departmental development budget controlled by non-project personnel, then effective control over the budget lies outside that of the manager. The FAO Agricultural Project Evaluation Study referred to in Chapter

3 indicates that shortage of funds is one of the main problems affecting implementation. As most public sector projects are not able to borrow from banks to cover short term financial shortfalls any failure to provide the agreed budgeted allocation results in the shut down of project activities. Typically, late or reduced allocations from the parent department leads to a cut-back of expenditure on essential equipment, material purchases and operating costs.

The problems associated with the shortage of project budget allocations, which may be due to entirely appropriate wider public expenditure controls, have, in recent years, significantly affected the way some development projects are implemented. Greater emphasis is now placed on the 'process' approach and on ways and means by which projects can be sustained through the generation of revenue flows out of project activities.

Financial management and accounting

While the project budget represents the implementation plan in financial terms, the actual operational aspects of handling the finance requires the development of appropriate organizational structures and systems. Most managers of development projects are not accountants, nor need they be. A basic understanding of accounting, however, may be of help to managers.

All projects of any size need an accounts section which supports three basic tasks:

1. Planning, control and recording of   ⎫   tasks of
      transactions   ⎬   accounts
2. presentation of information   ⎭   sections
3. financial decision-making   management function

The information resulting from the first two tasks allows management to undertake the third.

Planning, control and recording of transactions

These functions involve handling transactions such as payments for goods and services, materials, wages, etc, and handling receipts from sales, if any. This is a physical operation, but there is also a need for the planning of the transactions. Authority must be attained from project management for certain types of transactions and physical availability of funds must be arranged. Systems for authorizing payments and depositing funds in the project bank account must be established. The project budget is clearly of importance here as the approval of the project budget and the resultant release of funds to the project initiate the processes. The control function relates to matching physical records of transactions to approved budgets.

For many development projects this planning and control function can be a weak link in the project implementation process. It is at the point of carrying out transactions that the project is vulnerable to a variety of abuses. These can be simple ones such as the payment of wages to 'ghost' workers to more complex transactions in which, for instance, suppliers may provide the project with material or equipment below specification. The control function here is therefore of considerable significance. The degree to which this must be given attention by the managers will depend to a great extent on the nature of the project. Projects involving large stocks of materials which are purchased, stored and processed by the project require extensive investment into both physical and organizational systems to support these functions.

The management and control of stocks of materials and equipment is a major requirement of most development projects. While the control of physical money may be regulated through systems involving, for example, several signatures on each bank cheque, stocks are much less easily controlled. Pressure which can be exerted in many circumstances on project staff can be sufficient to cause the most well-meaning storekeeper to turn a blind eye to the movement of stocks. Even when there is no dishonesty involved, the pressure for 'urgent' use of materials can lead to borrowings and unrecorded transactions taking place.

Such movements of materials can very quickly give rise to chaotic management of stocks with negative impact on project implementation. Good stock control must involve at the minimum physically secure storage facilities and a systematic recording system. Each inward and outward transaction must be recorded. Other general rules on stock control, such as the desirability of old stocks being issued ahead of new stocks, on a 'first in first out' (FIFO) basis, are well documented by Austin (1984).

The bookkeeping system employed on the project will, to a considerable extent, depend on the its size and nature. Large public sector projects will generally be expected to implement systems normally followed by all the departments. Small projects may not require anything very sophisticated. Details of bookkeeping systems are beyond the scope of this text as they are the primary concern of project accountants rather than project managers. Readers seeking further guidance are referred to the comprehensive text by Coy (1982).

An important complementary function to bookkeeping and stock control is that of the internal audit. This involves the reconciliation of the records of transactions against the various project accounts and the physical stocks of money, material and equipment. The general purpose of the internal audit is to ensure that the books reflect the actual position of the project and not some paper concept. Internal audit sections play a vital role in projects and it is necessary for the unit to be in a position to report directly to the project manager. The internal auditors should not, however, be cast in the role of informers, but as quality controllers who are able to provide confidence in the system for managers and advice and support to the various account holders and stock controllers.

Presentation of information

This aspect of the financial accounting function involves the recording and utilization of the information provided under the planning and control function. The project accounts are likely to be of little value to management in their raw form. They may even be unintelligible to the non-accountant. It is vital, therefore, that a section of the project is responsible for aggregating the results of the various transactions, assessing their implications and summarizing these in regular reports which can be understood in clear terms by project management and project sponsors. Such reports will provide information of the progress on expenditure levels, include forecasts of future financial positions and indicate problem areas based on the records of transactions. This information will provide the basis of the decision making, the third element of the financial accounting function. Table 8.1 is a typical example of a summary project budget report drawn from a real project.

Table 8.1 Small enterprise development project: revised budget estimates 1991/92 Budget estimates 1992/93

Vote heading	1991/92 estimates	Half year actuals	Revised estimate	1992/93 estimates
	$	$	$	$
Staff loans	30 000	9 929	30 000	50 000
Capital investment				
construction	330 000	—	213 000	134 500
vehicles	25 000	—	53 564	—
furniture and equip.	7 571	25 244	76 495	41 225
computer	50 000	—	50 000	—
Recurrent costs				
personnel costs	599 323	228 472	489 713	896 199
travel costs	183 500	200 894	300 771	315 000
communications	71 000	33 061	76 120	82 000
property	132 572	72 946	121 552	149 540
legal and finance	16 600	18 024	23 600	23 600
interest on loans	—	—	—	20 000
miscellaneous	27 500	38 480	51 500	34 000
Technical assistance	385 896	202 309	442 309	432 000
Study fund	65 000	—	28 000	80 000
Training in-house	75 000	84 737	131 234	150 000
Total expenditure	1 998 962	914 096	2 087 858	2 408 064

Financial decision-making

In order for the managers to control projects effectively they must make the decisions that determine the progress towards meeting project objectives. Making decisions on financial issues is fundamental to this process. Managers often face problems involving decisions with respect to the allocation of financial resources between competing uses of the funds. It is therefore obvious that this decision-making should as far as possible be made on the basis of intelligible information derived from an accounting system that is well regulated by an internal audit. Figure 8.3 is a presentation of the management accounting function.

There is no single system of accounting for general use in projects and each project will require the establishment of a system appropriate to the nature of the project and the requirements of the project sponsors and funding agencies. However, all systems will need to incorporate the essential ingredients of a logical recording and checking function with a methodical reporting process. Failure to establish such a system will result in managers making decisions on the basis of inadequate information. This issue of developing management systems for decision-making is taken up more comprehensively in Chapter 9.

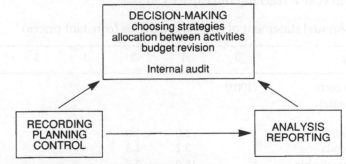

Fig. 8.3 Management accounting function

Investment appraisal

While project managers may not have any involvement in project formulation and appraisal it is nevertheless useful for them to have a clear understanding of the status of the project at the appraisal stage and particularly from the financial perspective. Investment appraisal is a very complex issue and is the subject matter of various books listed at the end of this chapter. For the purpose of this book the coverage of investment appraisal is restricted to the determination of a financial rate of return to a proposed investment which may be used for determining the relative efficiency of a project in providing a return on a particular investment. Ideally the purpose of investment appraisal is to enable a project to be ranked against others with a view to providing investors (public and private sector) with the opportunity to maximize the return on their available investment capital. In practice, decisions on investment are rarely made on the basis of a financial rate of return calculation alone. Many other considerations will need to be taken into account including economic, social, technical, risk and organizational issues. However, the basic elements of investment appraisal are discussed below, essentially in order to provide the reader with a link between project appraisal and project implementation and operation. It is often helpful for managers to know how their project has been planned and on what basis it was appraised. Some managers may also require the basic skills of project investment appraisal as carrying out such appraisal may be part of their terms of reference as project managers.

The discussion of investment appraisal and subsequently financial analysis in the next section will be undertaken through consideration of a simple illustrative case study, the details of which are outlined below.

Financial management case study: summary details

The case study development project involves an initial capital investment of $100,000. It is expected to operate over five years. All the capital investment is completed in the first year of the project, year 0. It begins operation in year 1, reaching full capacity in year 2.

Table 8.2 Annual statement of costs and benefits (constant prices)

Project year	0	1	2	3	4	5
Investment costs	100.0					
Working capital						
stocks						
materials		2.0	1.0			−3.0
final goods		3.2	1.3			−4.5
accounts receivable		15.0	7.5			−22.5
accounts payable		1.0	0.5			−1.5
Subtotal		19.2	9.3	0.0	0.0	−28.5
Operating costs						
labour		21.0	31.0	31.0	31.0	31.0
materials		12.0	18.0	18.0	18.0	18.0
overheads		5.0	5.0	5.0	5.0	5.0
Subtotal		38.0	54.0	54.0	54.0	54.0
Sales revenue		60.0	90.0	90.0	90.0	90.0
Net benefits	−100.0	2.8	26.7	36.0	36.0	64.5

Net present value at 10% Discount Rate (DR) = 18.1
Internal rate of return (IRR) = 15.2%

There are two categories of capital costs, the main investment costs (including plant, buildings and equipment) and working capital. The latter includes stocks of materials, spares, fuel, etc, and finished products. These are elements of 'physical' working capital. There is also a 'financial' element of working capital which reflects short-term credit for purchases and sales. (Working capital is discussed in greater detail later in this chapter.) At the end of the project's life working capital is recovered and is treated as a benefit to the project: in Table 8.2 it is shown as a negative cost.

Overhead costs cover items such as rent and administration. Operating costs include only those for material and labour for simplification. Revenue from sales begins during the first year of project operations.

Basic project resource flow statement and measures of project worth

Table 8.2 indicates the basic project statement in which the resources used and created by the project are shown as and when they are expected to be used or created, on a year-by-year basis. Project resource costs and revenue are measured in constant market prices at this stage and no account is taken of inflation or the actual financing of the project.

This statement has a number of purposes, including indicating what resources will be required to implement the project and when the benefits accruing to it will be realized. It is a description of the project expressed in resource value terms and is one of the major products of the formulation stage of the project cycle. The statement is used to determine the intrinsic 'worth' of the project in terms of its utilization and creation of resources. There are a number of different methods of measuring project worth.

The commonest traditional method is called the 'pay-back' method. The result of the pay-back method of project worth is the number of years that it takes for the value of the initial capital investment to be recovered from the net benefit flow as indicated in the basic project statement. For the case study project this would be four years.

The pay-back method gives a single indicator for decision-making, that is a number of years, which can be set against a predetermined norm or against that of other competing projects.

More sophisticated measures of project worth take into account the concept of 'the time value of money' and involve the technique of discounting. The concept assumes that value of money available 'now' is greater than at some time in the 'future'. The further into the future an amount will be available, the less its worth in today's terms.

In practical terms the concept is the reverse of the ability of money to earn interest while on deposit at a savings institution. To determine the future worth of an amount invested at a fixed interest rate the technique of compounding is used.

Future worth $\quad$ (FW) = $P(1 + r)^n$

eg $\qquad$ $FW = 100 (1 + 0.1)^1 = 100 + 10 = 110$

Where P (present value of investment) = 100 and r (the interest rate) = 10 per cent, and n is the number of years.

To determine the present worth of an amount available in the future the opposite process of 'discounting' is used so that

Present worth $\qquad$ $(P) = \dfrac{FW}{(1 + dr)^n}$

eg $\qquad$ $P = \dfrac{110}{(1 + 0.10)^1} = \dfrac{110}{1.10} = 100$

Where FW = value in the future and dr (the discount rate) = 10 per cent, and n = number of years

This process of expressing the value of amounts of money available some time in the future in terms of their present worth can also be applied to project cost and revenue flows.

By discounting the costs incurred by, and benefits accruing to, a project in each year using the same discount rate it will be possible to arrive at the total present worth of each and subsequently compare them. If the total discounted costs are subtracted from the discounted benefits, the net present value (NPV) of a project can be derived. The decision rule applicable to the NPV measure of project worth is that any project that has a positive NPV should be proceeded with as it demonstrates a greater 'earning' power than the other course of action, which is to invest in the alternative investment 'opportunity' expressed through the discount rate. In financial terms this might best be compared to the bank deposit rate of interest.

The problem with the NPV method of estimating project worth is that it provides an absolute value that cannot be used to compare one project with another in terms of their efficiency in generating benefits from a given amount of investment. This can be achieved through the determination of two other measures of project worth that both involve the use of the same discounting technique.

The benefit: cost (B:C) ratio is determined by comparing the present worth of the revenue and cost streams of a project and is normally expressed as a simple ratio. For the example project this is 1.2:1. There is a drawback to using the ratio for comparing alternative investment opportunities in that it can be estimated in two ways either by comparing the discounted gross benefits with gross costs or by comparing the discounted investment costs with the net benefit stream. The two methods provide a different answer and it is important to know which method has been used when comparing the B:C ratios of alternative projects.

A third measure of project worth involving the technique of discounting is the internal rate of return (IRR). When applied to this resource flow statement this is often referred to as the financial rate of return. The IRR is the discount rate at which the present worth of the benefit stream is equal to that of the cost stream. Put another way it is the discount rate at which the NPV is zero. To determine the IRR manually it is necessary to calculate the NPV of a project using a discount rate that provides a positive NPV and a rate that provides a negative NPV (ie a higher discount rate as the higher the discount rate the lower the worth of amounts occurring further into the future – usually the project benefits). The IRR can be determined from these two values through a process of arithmetical interpolation using the following formula or by plotting a graph of a series of NPV calculations.

$$\text{IRR} = \frac{\text{lower discount}}{\text{rate}} + \frac{\text{difference}}{\text{between}} \quad \frac{\text{NPV at lower discount rate}}{\text{absolute difference between NPVs at lower and upper discount rates}}$$

For the example case study project (Table 8.2), using the discount rates of 12% and 17%, the calculation is as follows:

$$\text{IRR} = 12 + 5 \left(\frac{(11.5)}{(11.5) - (-5.2)} \right) = 12 + 5 \left(\frac{(8.81)}{(8.81) - (-10.99)} \right)$$

$$= 12 + 5 \left(\frac{11.5}{16.7} \right) = 12 + 5 \left(\frac{8.81}{19.8} \right)$$

$$= 12 + 5 \ (0.69) = 12 + 5 \ (0.45)$$

$$= 12 + 3.45 = 12 + 2.25$$

$$= 15.45\% = 14.25\%$$

The IRR is normally rounded down to the nearest whole number, as the calculation is relatively imprecise. In this case it would be quoted at 14 per cent.

The advantage of the IRR is that it provides a single efficiency value. This can be compared with alternative investment opportunities in other projects and it avoids the necessity of having to determine a notional 'yardstick' discount rate used for estimating the NPV. The main drawback to the IRR is that in certain circumstances a project might have two IRRs, as in some cases where costs exceed benefits in later years of a project's life. It is, nevertheless, the measure of project worth most often used by major funding agencies such as the World Bank.

The purpose of calculating the 'worth' of any project is to assist in determining whether to go ahead with the project or not. However, it is important to keep in mind that the measures of project worth described above are only one variable taken into account in the decision-making process. For project managers other factors are likely to be of similar importance, even though it might be helpful for managers to know on what basis their project was appraised and what its worth prior to project start up was estimated to be. Some of these other factors might well be those that exert the greatest influence on the outcome of the measures of project worth estimation.

As indicated previously in this book, projects are planned and implemented in an uncertain environment. Changing estimates of individual cost and benefit items in the project resource flow statement (while keeping all the others the same) can be revealing in terms of which factors the project's worth is most 'sensitive' to. The results of such 'sensitivity analysis' are of particular interest to managers as it will provide some indication as to what factors might require particular managerial input and monitoring. For project managers the process of determining the financial worth of a project will not be particularly important. Most project managers become involved with a project only once the decision has been made to go ahead with it. The manager might usefully make use of the annual statement of costs and benefits, however, to determine forecast financial requirements and to assist in determining an overall financing plan for the project. In particular it will be necessary for the manager to adjust the statement to take account of inflation (if adequate information is available). This could make a big difference to the early years in which revenue is not being generated in significant amounts.

Financial analysis

As most development projects are intended to promote development objectives through the establishment of infrastructure or provision of non-commercial services, they do not readily lend themselves to the normal profitability analysis associated with commercial enterprises. The managers of such projects are not required to ensure that the project operates financially in the same way as a commercial business. There are other types of projects, however, which are designed to be financially viable enterprises. Managers of such projects are expected to behave more like entrepreneurs or business managers than public sector administrators. This increasingly applies in many developing countries where there is a growing need to ensure that the parastatal sector becomes less reliant on direct government budget subvention associated with economic reform programmes. Also many non-commercial projects are established to provide services and support to enterprises which clearly must be. Such client enterprises may be farms (large or small) or businesses in the industrial and service sectors. They may be privately owned businesses, co-operatives or parastatals. The managers of projects working to support such enterprises will need to have some understanding of the financial status of their clients in order to achieve the aims of the project satisfactorily.

The remainder of this chapter is, therefore, designed to provide project managers with the basics of financial analysis. It should give the reader sufficient understanding of the subject to interpret the status of a project through its financial statements. Readers requiring a more in-depth coverage of the financial analysis of projects are referred to the book on the subject published by the FAO Investment Centre (FAO 1990). Other useful books are listed at the end of the chapter.

For project management purposes there are three fundamental financial statements that can be used to determine the financial status of a project. These are

1. the trading and profit and loss account
2. the cash flow
3. the balance sheet

All three statements have specific purposes which are summarized below:

1. The trading and profit and loss account indicates the relative efficiency of the project operations (ie the utilization of project assets by management), represented in terms of income over expenditure over a given period of time.
2. The cash flow indicates the 'physical' flow of money through the project over a given period of time ie on a week-by-week, month-by-month or year-by-year basis. Only actual monetary transactions are taken into account in the cash flow. The cash flow indicates the liquidity of the project (its ability to meet cash requirements) and can be used to predict the return to owners' capital investment.
3. The balance sheet indicates the net worth and the nature of the project in financial terms at a 'specific point in time'. The net worth of an enterprise is indicated by the value of its assets. The balance sheet indicates how the assets have been funded from a mixture of owner's capital (shares), loans from financial institutions and retained profits from previous periods of operation.

During the implementation phase of a project, managers are for the most part concerned with the cash flow statement. They need to know that there is adequate finance available to meet the requirements of the project activities. During the operational stage, however, managers will be equally concerned with all three statements.

From whatever perspective that managers view a project, it is generally helpful for them to have a clear understanding of its overall financial status including that of the initial appraisal. The approach adopted for discussion of the financial statements in this book involves three sub-sections:

1. profitability, interest and depreciation
2. the cash flow statement and working capital
3. the balance sheet and ratio analysis.

Profitability, interest and depreciation

An analysis of the profitability of a project is undertaken from the determination and compilation of the profit and loss (P&L) account. This account has three main purposes:

1. to derive indicators of relative efficiency.
2. to determine the net profit to be incorporated in the balance sheet
3. to determine the tax liability of the project.

The profit and loss account compares costs against revenue over a defined accounting period, usually a year or half a year. The costs set against revenue include all costs incurred during the period and can generally be categorized as follows

1. direct operating costs: materials, labour, etc
2. overhead costs: management and administration
3. interest: loan interest and other charges
4. depreciation : capital equipment, plant, etc.

While there are no hard and fast rules for drawing up profit and loss accounts it is common for three measures of profitability to be derived through the account. The first is the gross trading profit, which is the difference between the revenue earned and the direct production costs. Deducting overhead costs, interest and depreciation from the trading profit provides the net profit before tax. The net profit after tax can be arrived at by deducting the tax due on the net profit at the appropriate rate. Normally any net losses from previous periods of trading can be set against the current tax liability. The profit retained after deducting tax can either be distributed to the owners of the enterprise, the shareholders, or used to increase the net worth of the enterprise as indicated in the balance sheet (see p.154–6).

The net profit before tax figure indicates the overall efficiency of the project in financial terms during the accounting period: it indicates whether the enterprise is capable of generating enough revenue to cover all the costs of production. However, it is important to note that the measures of profitability are not cash balances available to the project. This is due to the fact that the depreciation charge is not an actual financial transaction and that the profit and loss account does not take into account the repayment of loan principal, but only the 'cost' of borrowing in interest payments. The actual cash position of the project can be assessed only through the cash flow statement.

Most development project managers are not normally concerned with the efficiency of project operations as their main task is complete once the project has reached the operational stage. Nevertheless many may find it useful to have a clear understanding of the relative efficiency of an enterprise in financial terms. For example many rehabilitation projects or expansion projects involve both the management of the operations of an existing enterprise and investment activities related to the project. They may be inseparable activities. Process project managers may also be involved with an environment containing many client enterprises which the project is designed to influence. Managers of such projects will need to assess the relative efficiency of these enterprises through an examination of their profit and loss accounts.

For project managers the derivation of corporate tax liability is of secondary importance to ensuring efficient project operations. This can be looked at from a variety of perspectives, for example the comparison between sales revenue and the costs associated with direct production (such as materials, fuel and labour use), or the comparison of the administration costs to sales revenue. It is the function of operational management to improve these ratios in order to improve efficiency of production and therefore enterprise profitability.

Table 8.3 shows the profit and loss account for the case study project. Set against the revenue are the project operating costs, which give an indication of the trading profit. Overheads, interest and depreciation are deducted from the trading profit to determine the net profit figure. The amounts of depreciation and interest are calculated in each year through the schedules at Tables 8.4 and 8.5.

Table 8.3 Forecast profit and loss account (operating years)

Year	1	2	3	4	5
Revenue	60.0	90.0	90.0	90.0	90.0
Operating costs	33.0	49.0	49.0	49.0	49.0
Trading profit	27.0	41.0	41.0	41.0	41.0
Overheads	5.0	5.0	5.0	5.0	5.0
Depreciation	20.0	20.0	20.0	20.0	20.0
Interest	7.5	8.6	6.9	4.9	2.6
Net profit/loss	–5.5	7.4	9.1	11.1	13.4
Cumulative net profit	–5.5	1.9	11.0	22.1	35.4
Tax @ 40%		0.8	3.6	4.4	5.3
Net profit after tax	–5.5	6.6	5.5	6.7	8.0
Retained profit	–5.5	1.1	6.6	13.2	21.3

The information provided through the P&L account for this project includes the fact that, like most investment projects, it actually makes a loss during its first year of operation. The project is therefore not liable to pay any tax. In the second year, 2, however, the project moves into profit, but the previous year's loss has been set against the profit made in the second year for the purpose of tax assessment. After the second year of operations the net profit figure rises steadily over the life of the project.

145

If the case study project was to carry on for a sixth year the investment capital would need to be replaced and this would necessitate a new financing plan. The capital investment would have physically depreciated beyond its useful life. This depreciation factor has been taken into account in measuring project profitability and is discussed in more detail below.

Depreciation

It may take many years for a typical fixed asset such as a piece of machinery to wear out completely and become unusable. It would, therefore, be inappropriate to include the full cost of such equipment in any one year when attempting to measure project profitability, because the accounts would not accurately represent the true cost of production in those years. Instead, the cost of such assets should be spread over the life of the asset by deducting a proportion of their value in each accounting period.

Profit, as indicated above, is derived by deducting a cost element for fixed assets through depreciation. It should, however, be noted that depreciation is not a monetary cost. Fixed assets are generally paid for in full when they are purchased. Depreciation is a notional expense reflecting the cost of the assets over their operational life which does not reflect the actual purchase of the asset as a physical monetary transaction.

Depreciation of fixed assets can be calculated in one of three ways. Perhaps the most common is the 'straight line method'. In this case the economic life of the asset is estimated in years and the initial cost of the asset is divided by that number of years. In each accounting period an item for depreciation equal to the resulting value is deducted, being the same throughout the project's life. When the time is reached when the total amount of depreciation set against income is the same as the original cost, the item is said to have been 'written off'. Most assets, however, retain some vestigial 'salvage value' within the balance sheet. The straight line method of calculating depreciation is illustrated in Table 8.4.

Table 8.4 Depreciation schedule

Investment cost		100.00					
salvage value	0						
years' life	5						
Depreciation			20.0	20.0	20.0	20.0	20.0

An alternative method of calculating depreciation is the declining balance method. In this method assets are depreciated by a given percentage, usually related to the economic life of the asset, in each accounting period. Given that the percentage figure is applied to the remaining value of the asset after deducting the previous year's depreciation, it will be obvious that the amount of depreciation will be smaller in each successive year. Furthermore the asset will never totally be written off.

The setting of depreciation against income reduces profit and therefore tax liability. This helps to improve the cash flow position of enterprises, particularly in the early years when they tend to be short of cash. It is therefore usual for enterprises to use the first method if possible. The 'if possible' applies because often the method of allowing for depreciation is determined by government regulation. This is the third method of calculating depreciation in which the amount indicated in the accounts does not reflect actual physical wear and tear, but government investment policy. In most countries, commercial enterprises, public companies and projects are required to publish accounts by law and some consistency is necessary. Furthermore allowances for depreciation are often structured to encourage investment. This can be achieved by allowing enterprises to reduce their tax liability during the early years of investment by depreciating assets more rapidly in the accounts.

A common feature affecting the calculation of depreciation in many developing countries is that rather than depreciating in financial terms, assets actually 'appreciate' in value over their lives up to a point at which they are no longer physically serviceable. This is as a result of rapid inflation and a general shortage of such assets. Much the same applies to assets such as land, where the value of land rises over time and is normally never worn out. Accounting procedures in such circumstances can become complicated. Project managers and accountants will need to be aware of the practices governing such matters in their own national legal framework.

Interest, loan repayment and the financing plan

Prior to the implementation of any project consideration needs to be given to the way the project is to be financed. A financing plan will be needed to enable the costs of the project to be met, particularly during the early years of establishment and implementation, before it generates enough money from its own operations to meet these requirements. A financing plan is usually the subject of much negotiation between project owners, sponsors, potential investors and financial institutions. The purpose of the financing plan must be to ensure adequate liquidity for project management to achieve project objectives. This adequacy can be assessed only through the process of forecasting the likely outcome of project operations and is therefore liable to a high degree of uncertainty. A basis for beginning the process of deriving a financial plan would be consideration of the basic resource flow statement, updated prior to project implementation and adjusted for inflation over the first few years of the life of the project.

If the project financing plan includes borrowing money from institutions the interest charges levied must be determined in order to estimate project profitablity. The repayment of loan principal is taken into account in the cash flow statement. (The cash flow statement can also be used to measure the return on capital invested in a project as discussed later in this chapter.)

The calculation of the interest figure in each year of the loan repayment period will depend on the method of loan repayment applied. One of the commonest methods is to repay the lender in equal instalments during each repayment period consisting of varying proportions of interest and principal. To determine the instalment it is necessary to perform an arithmetical function involving the total loan outstanding at the beginning of the repayment period, the number of years over which the loan is to be paid off and the interest rate. Fortunately standard numbers apply to the latter two components of the function, known as 'capital recovery factors'. These are readily available in tables denoting years and interest rates and are incorporated in most computer spreadsheets.

Alternative methods of loan repayment include dividing the outstanding principal by the number of years of repayment and calculating the interest on the basis of the declining balance. This makes calculating interest easier, but involves the enterprise having to make unequal total loan repayments in each year. An even simpler method of loan repayment is to pay off all the loan at the end of the repayment period in a lump sum. Interest payments under this method will remain the same for each year until the loan is paid off. Few commercial lenders would make loans on the latter basis.

The financing plan for the case study project involves a bank loan of 50 per cent of the value of the initial capital investment (ie $50 000) the other half of the cost being met by the project sponsors as equity investment. The bank loan bears interest at 15 per cent and is to be repaid over a five-year period as from the end of the first year of operations. The loan is to be repaid in equal instalments of varying proportions of interest and principal. The main advantages of this method of loan repayment to the borrower, besides the convenience of being able to pay an equal instalment each year, is that a relatively high proportion of the early repayments is made up of interest. This is significant in projects whose profits are subject to taxation for the same reasons as discussed in connection with depreciation.

Table 8.5 illustrates the method of calculating loan repayment using the equal instalment method. Note that in this simplified example there is a one-year grace period on the loan repayment. A grace period is the time between drawing down a loan (the act of actually using the funds negotiated with the lender and the point in time at which interest charges become payable) and the beginning of the repayment period. During a grace period interest only would be charged, either payable as it is incurred or added to the principal already outstanding, as in the example project. The latter process is known as 'rolling up' or 'capitalizing' the interest. Once interest incurred is added to the principal, interest becomes chargeable on both the original principal and the interest added. The mechanics of deriving the amount of annual instalments due under the method of equal annual repayments involves the use of predetermined 'factors' for a given interest rate and number of years over which the loan is to be repaid. These 'capital recovery factors' are readily available from financial institutions and are included, along with other useful project analysis factors, in *Compounding and discounting tables for project evaluation* published by the Economic Development Institute of the World Bank (Gittinger 1973).

Table 8.5 Loan interest and repayment schedule

Year		1	2	3	4	5
Loan owing year start		50.0	57.5	46.0	32.7	17.5
Interest rate %	15					
Unpaid interest		7.5				
Repayment period years	4					
Repayments		0.0	20.1	20.1	20.1	20.1
interest		0.0	8.6	6.9	4.9	2.6
principal		0.0	11.5	13.2	15.2	17.5
Balance at year end		57.5	46.0	32.7	17.5	0.0

In summarizing this section on profitability it can be stated that the trading and profit and loss account is used to determine the efficiency of project operations by comparing revenue (or benefits) against costs. Profit is a conceptual figure. It is not to be confused with a cash surplus. Spreading the costs of capital items over their physical life through depreciation, rather than when they are actually paid for, and taking into account only loan interest payments (which are a project cost) and not repayment of loan principal (which is not a project cost) accounts for the difference.

It is entirely conceivable that during a period in which an enterprise is making very good profits it may run out of cash and therefore cease trading. This is because during the same period loan principal repayments, which are not included in the profit and loss account, may fall due. If these repayments are higher than the profit, plus the allowance for depreciation and any cash reserves carried forward, then the enterprise will have run out of funds to meet the necessary repayments. Unless management can renegotiate the terms of the loan repayment or borrow more money to finance the deficit, then the enterprise will have to cease trading and liquidate its assets to pay off the debt. This scenario is a familiar feature of the corporate and commercial environment and applies as equally to small scale village-level enterprises as to large commercial companies.

The cash flow statement and working capital

The cash flow statement indicates the inflow and outflow of physical cash within the project over a specific period. Cash flows can be drawn up on a daily, weekly, monthly, quarterly or annual basis. For project management purposes the forecast cash flow drawn up on a monthly basis over the forthcoming quarter may be the most valuable. The cash flow simply indicates whether or not the project is able to meet its expenditure requirements. As stated earlier in this text one of the most common problems of development projects is an acute shortage of funds to finance planned expenditure. Drawing up a forecast cash flow is, therefore, a fundamental management task if early warning signs of financial constraints are to be detected.

The elements of the cash flow are the inflows of money into the project and the outflows. Inflows include revenue from sales, loans received and drawn down, equity capital contributed from shareholders, loan repayments made to the project (if any), any grants received plus any cash balance brought forward from the previous accounting period. These added together constitute the total inflow of funds.

The outflows include expenditure on production inputs such as materials, labour, parts, and so on (note that this is expenditure on such items, not their actual use and in normal circumstances payment for such items is made some time after actual delivery and use), overhead items such as staff salaries, rents, etc, tax payments made, loan repayments including interest and principal, and expenditure on new capital items such as vehicles, plant and equipment. The difference between the total inflow and total outflow is the net cash balance, which is carried forward to the next accounting period (see Table 8.6).

Table 8.6 Forecast cash flow for financial planning

Year	0	1	2	3	4	5
Cash inflow						
equity	50.0	0.0				
loans	50.0					
revenue	0.0	60.0	90.0	90.0	90.0	90.0
Annual inflow	100.0	60.0	90.0	90.0	90.0	90.0
Cash outflow						
investment	100.0					
working capital	0.0	19.2	9.3	0.0	0.0	−28.5
operating costs	0.0	33.0	49.0	49.0	49.0	49.0
overheads		5.0	5.0	5.0	5.0	5.0
loan interest	0.0	0.0	8.6	6.9	4.9	2.6
loan principal	0.0	0.0	11.5	13.2	15.2	17.5
tax	0.0	0.0	0.0	0.8	3.6	4.4
Annual outflow	100.0	57.2	83.4	74.9	77.7	50.0
Annual cash flow balance	0.0	2.8	6.6	15.2	12.3	40.0
Cumulative cash flow	0.0	2.8	9.4	24.5	36.8	76.8
Equity cash flow	−50.0	2.8	6.6	15.2	12.3	40.0
Return to equity 11.4%						

The important point to note with respect to the net cash balance is that the balance must be positive for the project actually to achieve the planned expenditure. A negative cash balance indicates that the project does not have sufficient funds to finance projected expenditure. This does not mean that the project must necessarily stop. It is the task of management to determine whether some expenditure could be delayed or reduced in order to meet the financial constraints, consistent with causing least damage to the achievement of project objectives. Alternatively ways may be found to increase revenue or if this is limited other options include

1. asking shareholders for more equity capital
2. asking lenders for increased loan funds
3. asking lenders to defer loan repayment
4. requesting suppliers to extend the period of credit given on the supply of goods or services.

Clearly any managerial action to meet a deficit in project finance will have implications and in some cases these may be significant. Short-term contingency action may be able to assist the project to stay on track, but if the cash flow problems are chronic, requiring major adjustments in project financing, then management will need to ensure that the project sponsors are fully aware of the situation. The failure of project management to foresee cash flow constraints often leads to project failure. Forecasts can be wrong and given the uncertainties in forecasting prices and quantities it would be surprising if cash flow forecasts proved any more reliable than other types of forecasts. The need to apply sensitivity analysis in order to identify the most important factors in the cash flow is therefore another key task of financial management.

A further use of the project cash flow of general interest to potential investors or lenders to a project, is the return to equity figure that can be calculated from the projected project cash flow. In the earlier sections on investment appraisal the FRR was calculated. However, this is simply an indicator of financial earning power of the project in 'resource' terms and it does not refer to monetary returns. The return to equity on the other hand specifically refers to the cash return accruing to the holders of equity (ie the owners' own funds invested) funds invested in the project. The return to equity is the rate of return to the equity cash flow as illustrated in Table 8.6. It is this figure that is used as a basis for decision-making by investors as it takes into account the financing plan and the cost of financing the project through borrowing, which basic investment appraisal does not.

Working capital

It has been suggested earlier in this chapter that for project managers responsible for the implementation phase of the project, the most important tasks with respect to financial management are the drawing up of budgets and the preparation of a cash flow statement. These indicate what expenditure is expected to be incurred and whether there will be adequate cash resources available to meet this expenditure.

While this may be so, it is almost invariably the case that development project managers operate in a resource-constrained environment where the emphasis is on the achievement of project objectives at least cost. There is usually a trade off between achieving targets on time and the cost implications. Most things can be achieved more quickly if more money is allocated to the task. In modernized economies such things as spare parts and material inputs are readily available for purchase and immediate delivery. Almost by definition this is not the case in a developing economy. This creates difficulties for management and has implications for project finance. In particular it relates to the management of 'working capital'.

Working capital is a phrase that can easily be misunderstood due to its use in everyday language. But from a management accounting perspective it has a precise meaning best described through an illustration of how it is made up. One of the main components is that of stocks of materials. These could be materials used in the productive process or for maintaining operations such as fuel, spare parts, raw materials, and so on. No enterprise can operate without such stocks. They are used up during day-to-day operations and are generally replaced on a regular basis. A volume of such items will remain unused at the end of each accounting period and appear as current assets in the balance sheet (readers may wish to refer to the next section on the balance sheet here if they are unfamiliar with this financial statement). Likewise it would be unusual for any productive enterprise to be able to sell off all its output on the day it is produced. Therefore, at any one time there are likely to be stocks of completed and partially completed outputs in the project which feature in the balance sheet as current assets.

In addition to physical stocks, the value of which can be calculated with relative ease, any project will need some cash in hand to service the day-to-day requirements of running the project. This will be used up continually for various purposes and continually replaced. Such cash in hand is also treated as a current asset.

While stocks and cash in hand are the physical elements of working capital they do not necessarily represent the full picture of financial working capital. This is because in reality payments for goods and services bought and sold by the enterprise are rarely carried out simultaneously with the physical transactions. Commercial activity can be continued only if suppliers of goods and services are prepared to supply such items and wait some time for actual payment at a later date. Suppose this delay in the payment for goods and services was not reflected in the project accounts and if relatively large stocks of material inputs were recorded as current assets which had yet to be paid for, the accounts would give an inflated impression of the true asset value of the enterprise. Likewise, if the enterprise had provided goods and services for which it had not yet been paid, and this was not recorded in the accounts, then this would under-estimate the true worth of the enterprise at that particular time.

In order, therefore, to make adjustments to reflect these credit transactions the items of debtors and creditors are taken into account in the asset and liability items. Debtors represent the amount owed over the short term to the enterprise; creditors represent the amount owing over the short term by the enterprise. Creditors for this purpose covering the short term are treated as current liabilities; they are not to be confused with long-term creditors (lenders) which feature in the long-term liability section of the balance sheet.

In summary working capital can be portrayed as

working capital = stocks + cash + debtors − creditors

This is the same as the net current asset figure in the balance sheet (see next section).

A very important point with respect to working capital is that, like capital expenditure, it needs to be financed. This is sometimes confusing because the common usage of the term working capital refers to the cash component of the item which is seen by some as a 'source' of finance. In fact when negotiating project finance the size of the working capital can be quite significant and cannot be ignored. For some projects, such as agricultural production projects, the working capital requirement can be of greater size and significance than the investment in fixed capital. This is because farming uses large amounts of inputs which need to be tied up for several months before there is any revenue to pay for them.

Working capital is generally paid for by short-term borrowing. Banks and equity holders are usually willing to lend long term to projects only for the purchase of capital items, such as fixed assets which they can use as security for their investment. Financing working capital can be expensive in that the interest rates on short-term borrowing can be very high. Consequently an important task of project management is to reduce this interest burden as far as possible, thereby not only increasing profit, but also, just as importantly, easing the burden on the liquidity position of the project as shown in the cash flow.

Minimizing the interest charge incurred through the funding of working capital can be achieved by a number of methods. One is to extend the period of suppliers' credit. Negotiating an extra few weeks' leeway in payment for goods provided may prove crucial. Alternatively it may be possible to insist that purchasers of goods and services from the project pay for them as soon as possible. Another strategy would be to minimize the requirement for stock holding. In some industries in modernized economies, for example the motor manufacturing industry, sophisticated methods have been deployed to minimize the requirement for working capital. 'Just in time' delivery systems have been developed in which materials and components required for production are delivered to the assembly plants virtually on the day they are required. Such an approach minimizes the need for expensive storage facilities and the need to hold large stock holdings.

In the context of projects in developing countries the priority with respect to management of working capital may not necessarily be the need to minimize its cost. The priority is often to make sure that the working capital component is realistic in terms of keeping the project on schedule. The availability of materials such as fuel and spare parts may be so erratic that the only way to ensure continuity of project operations is to hold very large stocks. Unfortunately funding agencies are often reluctant to finance such large holdings of assets that, by definition, are mobile, readily exchanged into cash and are in short supply. Problems of asset security and abuse of project resources may also influence the level of working capital in terms of stocks held.

The task of managing project working capital is, therefore, not only one of reaching an acceptable trade off between the cost of its financing and the requirement to ensure smooth implementation, but also one of ensuring adequate management of the physical materials and cash.

The management of working capital has considerable implications for development projects. All too frequently projects such as livestock enterprises are found in the situation where food stocks have run out due to local shortages, insufficient purchases of initial stocks, or use of stocks for other purposes. Such experiences indicate a general lack of understanding by managers of the importance of working capital management.

The balance sheet and ratio analysis

A balance sheet is a financial statement showing where money came from (sources of funds or liabilities) and how it was spent (assets). These are related to each other in the following way. The project (or company/cooperative, etc) consists of assets financed by long- and short- term loans. When these are subtracted from the value of the assets the remaining figure represents the shareholders' stake in the enterprise (equity capital). The shareholders' element normally consists of capital subscribed by the shareholders (ie originally invested in the enterprise) plus retained profits.

There are two ways of setting out a balance sheet, horizontally and vertically. The vertical layout for the case study is shown in Table 8.7. While the two layouts provide the same result in terms of the balancing items the vertical layout has the advantage that the difference between the 'current' assets and liabilities are clearly shown. The difference between these two items is an important indicator of the liquidity position of the project, which for management purposes is of particular value. Whatever approach is used in laying out the balance sheet the total assets and liability figures must, by definition, balance.

Given that loans and credits are clearly measurable sums in the balance sheet, it is the value of shareholders' stake which varies relative to the change in the value of assets. If the enterprise performs well and makes a profit, this boosts the shareholders' stake. If it performs badly and makes losses, the value of the shareholders' stake is diminished. If it consistently makes losses so that the total assets become almost equal to the value of the outstanding loans, then the value of the shareholders' stake will be approaching zero. The lenders may demand that the enterprise be wound up so that the assets can be sold off to repay the outstanding loans.

Table 8.7 Forecast closing balance sheet

Operating year	1	2	3	4	5
Fixed assets (less depreciation)	80.0	60.0	40.0	20.0	0.0
Current assets stocks					
materials	2.0	3.0	3.0	3.0	3.0
finished goods	3.2	4.5	4.5	4.5	0.0
cash	2.8	9.4	24.5	36.8	76.8
accounts receivable	15.0	22.5	22.5	22.5	0.0
Current liabilities					
accounts payable	1.0	1.5	1.5	1.5	0.0
loan due payment	11.5	13.2	15.1	17.3	0.0
tax due		0.8	3.6	4.4	5.3
Net current assets	10.5	23.9	34.1	43.2	71.4
Total net assets	90.5	83.9	74.1	63.2	71.4
Liabilities					
equity	50.0	50.0	50.0	50.0	50.0
long-term loan	46.0	32.7	17.5	0.0	0.0
retained profit/loss	−5.5	1.1	6.6	13.2	21.4
Total liabilities	90.5	83.9	74.1	63.2	71.4

Information presented in the balance sheet is drawn from a variety of sources to present a picture of the project at a particular point in time. As the project continues to be implemented and operated the values of the items in the balance sheet change. (For instance, the value of equipment reduces through depreciation, loans are paid off, and so on.) The purpose of the balance sheet is primarily to enable shareholders and lenders to review the status of the enterprise at a given point in time with a view to determining the relative security of their investments.

For the purposes of project management the balance sheet has two particular functions. The first involves the ratio between the proportions of total assets owned by shareholders and the proportion of assets due to lenders to the project. This ratio is commonly referred to as the 'gearing ratio'. A high gearing implies that the ratio of borrowed money to shareholders' stake in the project is relatively high. A project which is almost totally funded by the owners of the project has, by contrast, a very low gearing ratio.

This ratio is of particular interest to prospective financing agencies and donors. Clearly the more highly geared a project is then the more closely prospective funding agents would wish to examine the viability of the project. Generally lenders want to see the owners of the project having a significant stake in it. If most of the assets are financed by borrowing then the greater the risk of the lenders losing money if the project fails. Many financial institutions have regulations that will not allow them to lend more than a given percentage of the total asset value of an enterprise. More commonly such institutions limit their funding of a project to the 'fixed' assets which, in the event of the project failing, they would at least be able to sell and recover their value. It is always the case that the creditors of an enterprise make first call on the enterprise assets in the event of failure and it is therefore the owners (or shareholders) that carry the greatest risk.

Another main item of information that can be derived from the balance sheet is the ratio between the value of fixed assets and the net current assets. The difference between the fixed and current assets lies not in the fact that fixed means literally fixed, but in the relative ease with which each can be realized as cash. Fixed assets such as buildings and plant and equipment take time to realize as cash as they take time to sell. Once any part of the fixed assets of an enterprise have been liquidated, without being immediately replaced, the enterprise is no longer capable of operating to its specification, so the asset configuration has been changed.

This is not the case with current assets, which consist mainly of cash, stocks of material inputs and completed outputs. These items can readily be sold and converted to cash. Furthermore, even if they were all liquidated in this way the intrinsic operational capacity of the enterprise will remain intact, that is the overall asset configuration will not have been affected (although its capacity to operate in the immediate short term will have been affected).

By looking at the ratio between the net current assets (or 'working capital' as referred to above) and the fixed assets it is possible to determine to some extent the nature of the project or enterprise. Major industrial projects, such as an iron smelter, may have very large fixed assets in the form of plant and equipment, but a relatively low level of net current assets. Conversely a marketing project involved with purchasing, storing and processing agricultural produce may have a great deal of money tied up in current assets, particularly in the form of stored produce, but relatively little by way of fixed assets. In addition to these ratios there are a range of others that can be determined from the balance sheet, and in combination with the other statements, provide indicators of enterprise performance over time. Detailed coverage of these can be found in the texts referred to at the end of this chapter.

Summary

The purpose of this chapter has been to provide existing and prospective project managers with knowledge of the basics of project finance. The main elements from a management perspective are the role and mechanics of budgeting and the meaning and purpose of the financial status of a project as expressed through the financial statements. The chapter also included a brief summary of the methodology of project investment appraisal to allow managers to establish a relationship between the appraisal stage of the project cycle and project implementation.

Clearly financial management and accounting are major study areas in themselves. It is not possible in a book such as this to treat the subjects in great depth. It is expected, however, that managers of projects will be able to use the concepts and techniques covered in this chapter in conjunction with those covered in Chapters 6 and 9 to develop appropriate systems for managing the finances of a project through the project establishment phase.

References and further reading

Austin V 1984 *Rural project management: a handbook for students and practitioners.* London. Batsford Academic and Educational London.

Coy D V 1982 *Accounting and finance for managers in tropical agriculture.* Intermediate Tropical Agriculture Series, London, Longman.

Gittinger J P (ed.) 1973 *Compounding & discounting tables for project evaluation.* Washington DC, EDI, IBRD.

FAO 1990 *Design of agricultural investment project lessons from experience.* Rome, UNFAO.

MacArthur J D 1988 *Appraisal of projects in developing countries: a guide for economists.* London, ODA.

Price Gittinger J 1982 *Economic analysis of agricultural projects.* Economic Development Institute (EDI) Baltimore, Md, Johns Hopkins University Press.

Selvavinayagam K 1991 *Financial Analysis in Agricultural Project Preparation.* FAO Investment Centre Technical Paper No. 8, Rome, FAO.

CHAPTER 9

Project management systems

Introduction

This chapter focuses on the three key systems of project management:

1. the monitoring system
2. the control system
3. the information system.

These are linked to one another and are the basis of the project monitoring and control cycle shown in Figure 9.1.

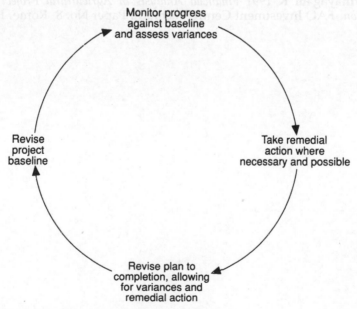

Monitor progress
against baseline
and assess variances

Take remedial
action where
necessary and possible

Revise plan to
completion, allowing
for variances and
remedial action

Revise
project
baseline

Fig. 9.1 The monitoring and control cycle

The essence of a project monitoring system is the continuous comparison of the actual situation against the plan, in relation to physical progress, financial expenditure and quality (the fitness of the assets created by the project for the purpose for which they were intended). When the monitoring system shows a deviation between the planned progress and the actual progress (known as 'variance'), project managers need to decide whether remedial action is possible or necessary.

Once the monitoring system has been established, managers will need to institute a control system, which takes action speedily and effectively when deviations from the plan are noted. It is the essence of the manager's job to decide what corrective action is necessary in any particular situation, to authorize and implement the proposed changes and to adjust the control baseline so that appropriate monitoring can continue. This may well require a considerable degree of communication and interpersonal skills, as well as technical modelling and analysis.

The function of the information system (often called the PMIS – the project management information system) is to provide managers with the information on which they can take timely action in pursuit of project activities. The information (and communication) system underpins the monitoring and control systems. Information is passed from the project site to the managers, who use it to monitor project progress against plan. If deviations have occurred and modifications to the plan are required, these are instituted through the control system, and the consequent decisions are passed back to the team at the project site using the communication system. In fact the PMIS is likely to provide the media for transmission of information and communication in both directions (from site to management and from management to site).

Desirable features of management systems

Without doubt, it is most important that the managers concerned should feel an 'ownership' of the management system of their project. Monitoring and control are, or should be, internal processes: they are carried out by managers, for managers, as part of the effective implementation of projects. Those involved must look on the management systems as being an assistance to them in the carrying out of their own duties, rather than as something required by others for different purposes. There is a danger with many types of comprehensive systems proposed at national level and based on electronic technology that managers will feel that they have performed their duties in relation to monitoring when they have transmitted the information to the computer operator, and it has then been passed back to the central ministry. Too often, in public sector administrations, such monitoring systems are in fact a means of centralized checking and recording, particularly focusing on financial performance. For instance, a group of Indonesian project managers in the irrigation sector with whom one of the authors worked on devising a PMIS, all proposed various mechanisms whereby information on their project could be transmitted to the head office in Jakarta, rather than systems to help them manage their projects more efficiently. A similar approach was observed in the Rehabilitation Programme in Sri Lanka, where the proposed monitoring system was in fact a very detailed system for recording expenditures at project sites. Some form of centralized control is inevitable where public sector accountability is important but managers should endeavour to do all they can to institute a PMIS which helps themselves to manage.

Indeed the relationship between a project-based PMIS and a national-level implementation monitoring system is one that needs a considerable degree of thought. There is an important role for a national monitoring system to play but this is in the identification and rectification of national policy constraints and the removal, as far as possible, of external hindrances, rather than in the control and supervision of individual project managers, which is the ostensible purpose for which many national systems are established. The project-based PMIS is an important part of the overall national system but there may well be structural differences between them. It is most unlikely that an information system could be devised which would fit all levels of the hierarchy. 'Ownership' of a management system is felt more positively when the managers themselves participate in its establishment, thus reinforcing the idea that imposition of centrally planned systems is not, in general, a fruitful basis for the design of a PMIS.

The function of management systems is to assist in the effective implementation of projects. The next most important feature is that they should therefore be at all times forward-looking, using past information only to the extent that it helps in forecasting future events and suggesting possible courses of action. For this reason, only essential information should be requested, that is information which has a direct bearing on the progress of the project. Additional but inessential data (for instance, personnel or perhaps marketing information, for a particular project under implementation) should not be requested on a routine basis. Concentration on essential data will also ensure that the systems are relatively simple and are therefore cheap and do not require additional resources of their own to operate.

While all monitoring data need to be processed to provide information, managers should use that information to guide them to take action only where necessary. The data will provide information on which it will be possible to compare actual progress against the plan. Areas of concern are those where there are substantial deviations from the plan: the manager will then be concerned to assess the implications of these deviations. In some cases, perhaps, these may not be important. In others, they will be seen vitally to affect other components of the programme and the managers will need to consider remedial action. In management terms, this is called 'management by exception'. Managers do not need to take action on monitoring information except in those situations where the information shows substantial deviations between planned and actual progress.

Other desirable features of management systems are that they should be timely and accurate. There is clearly a trade off between these two and the aim must be an optimum combination of them. For management purposes, it is important that data become available in reasonable time to make subsequent action effective. If additional accuracy requires a delay of a month or two, then serious consideration should be given to whether the reduced level of accuracy of immediate data would still permit effective and informed decisions.

With questions of accuracy go considerations of verification. It will generally be appropriate to assume that the information being supplied is genuine and factual and other measures to obtain verification need be taken only in exceptional circumstances. However, the value of site visits should not be underestimated as a means of providing general back-up to the monitoring data. It should, for instance, be possible for the manager to gain a reasonable estimate of general project progress and the time required to complete particular components during a site visit, which can be correlated with the monitoring data supplied. Site visits also provide an opportunity for controlling the direction and quality of the project output, as well as playing an important role in fostering the relationships between the manager and site supervisors.

Project monitoring systems

In discussing the basic content of a monitoring system, we must start off by emphasizing the distinction between those systems which are designed strictly to assist in the efficient implementation of a project, compared with those which may be concerned with the operational stage. There is, first, the requirement for project monitoring, concerned with delivery of inputs to the project and provision of outputs by it (using the terminology of the project framework). This type of monitoring is of particular concern to project managers and forms the main topic of this chapter: it is required at all stages of project development, and especially during implementation. Once project implementation has been completed and the project has been commissioned or established, there will be a need for ongoing or operational monitoring, to determine the impacts and effects of the facilities or assets created by the project (again using the terminology of the project framework). Operational monitoring will be concerned, for instance, to measure the utilization of health facilities or the level of agricultural production on a continuing basis. Measurement of effects and impacts is also the key element of evaluation. It is worth emphasizing the distinction between 'monitoring' and 'evaluation', though the two are often referred to together. Monitoring is an internal management activity, undertaken during the progress of project implementation, in order to promote its success.

Evaluation is the process of investigating the success of a project particularly related to its purposes, and perhaps the goals of the programme of which it forms part. It asks the question 'did this project succeed in achieving its purpose?' (for instance, opening up a rural region to urban markets), which is a different question from 'is this project on schedule to achieve its planned output?', (the construction of 20 km of road). The latter question is part of monitoring and would be undertaken by the manager/ implementor. Evaluation, by contrast, is normally an external activity, that is one carried out by those outside the project organization itself.

For project monitoring, key information concerns

1. measurement of physical progress
2. measurement of financial progress
3. quality control, and the fitness of the project outputs for their intended purposes
4. other information specific to the project, including, for instance, environmental aspects if these are of particular concern.

Essentially, project monitoring consists of developing a plan in each of these areas and measuring progress against the plan. In doing this a number of subsidiary systems will be brought into play, including for instance the budgeting, cost control and accounting systems described in Chapter 8. These are the essential elements on which the management systems are built, but it is not within the scope of a book of this nature to examine them in detail, particularly as such financial systems for public sector projects are likely to be legislated by central government. The monitoring system must enable the easy identification of variances, that is deviations between the plan and the actual situation, and the reason and implications for such variances (note that these variances are quite distinct from the use of the word 'variance' in statistical analysis).

The monitoring system depends on the definition of the project plan through the identification of project targets, to be achieved at a particular time. These targets describe the quantified objectives of the project at the particular time. It is important to appreciate that, for effective monitoring, it is necessary to set physical targets for outputs as well as financial targets (budgets) for inputs. Often targets appear to be in financial terms, either in the form of budgets or of annual targets expressed as a percentage of the total estimated cost for all the activities to be completed in a particular project. Such budgets or targets are an essential element of control of financial expenditure, but, in addition, it is necessary to identify simple physical targets for each activity within the project.

Generally, progress can be measured directly against targets, particularly at the level of project inputs and outputs. The most important input is financial resources, for which costs expended can be compared directly with the budget. Other inputs such as human-power can usually be similarly quantified and measured. At the output level, quantified targets can often be defined, for example, 20 km of road to be constructed during the year. Here again, progress can be measured directly against these targets. Even in the case of institutional and non-physical projects, the definition of project outputs should not present great difficulties since its output targets should be capable of relative precise definition (number of clinics built and equipped, number of health workers trained, development of appropriate village-level training courses etc).

Monitoring physical progress

Monitoring physical progress is a vital element of successful project implementation. It is generally accepted that projects which are implemented on time have a much better chance of being implemented within budget than those which suffer delays. Physical progress monitoring should therefore be directed to assisting the managers and owners of the project in keeping a check on whether activities in the project are up to schedule. If they are not, managers need to be able to assess how significant the delay is, and whether remedial action needs to be taken. Managing physical progress can be likened to managing time.

The simplest method of physical progress monitoring is by means of project milestones. Milestones are significant events in the implementation programme, for example, the event marking the completion of the various critical activities. Completion of the tendering process and the award of contract is often a very important milestone for resource-intensive projects with major physical components. Failure to achieve milestones by the appropriate date signals to the management that problems are likely to occur in the future, and that remedial action may be necessary.

The number and type of milestones used for project monitoring will depend on the level of management that wishes to use them. Top management will be concerned with a few key milestones which will indicate progress on the project as a whole. These milestones are likely to represent important dates in major activities or complete components of the project. Middle management will be concerned with more milestones, at more regular intervals, and relating only to that part of the project for which they are responsible. Line supervisors will be concerned with the achievement of detailed milestones, at frequent intervals, for one specific activity. Identifying appropriate milestones is a matter of judgement. If, however, critical path methods are used to plan project implementation, the project network and time charts are useful aids in this respect. When identified, the milestones can be marked up: this is commonly done on the bar chart.

An effective monitoring aid using milestones is the milestone chart (Table 9.1). This shows, for each project milestone, the originally planned date, the revised planned date, and the achieved date. Such a chart, besides showing the original plan, allows the establishment of a revised plan when deviations make this necessary, as well as keeping a record of actual progress which may be very useful in subsequent evaluations of project implementation.

Monitoring using milestones concerns the occurrence of key events and only signals when problems have already happened. It does not help managers to anticipate problems. To do this it is necessary to use measures of physical progress which are regularly reported during the activity and which help the project manager to look forward to assess whether future milestones will be achieved on time or not.

Table 9.1 Milestone chart

Milestone	Original plan	Revised plan	Date achieved
Award of tenders	April 1990	—	April 1990
Completion of buildings and physical facilities	Sept 1990	Dec 1990	
Completion of staff recruitment and training	Oct 1990	Jan 1991	
Completion of pilot phase	Jan 1991	April 1991	

The chart shows data entered in late 1990, at a point when delays to the physical facilities have caused delays to the whole project, and required a revised plan and completion date.

In defining measures of physical progress three situations may be distinguished. In the first (and simplest), the output of the activity can be quantified as a single number, for example, the number of wells for a water supply project. In this case both the physical target and the physical progress can be expressed in terms of this single number.

If progress cannot be so quantified, the next best measure is 'percentage completed'. However, there is a difficulty in using measures of progress expressed in percentage terms. What does this percentage mean? What does it tell us about the work and the time remaining on the activity? Many projects seem to be making reasonable progress until the last 5 per cent. In other cases deviation from the original schedule are so great as to make the initial estimates of duration meaningless.

The second situation that can therefore be distinguished is when the output of the activity can be measured and valued – for example, in the construction or repair of buildings, roads and other physical works. In this case physical targets and progress should be expressed in terms of:

$$\frac{\text{Value of work done}}{\text{Total value of work planned}} \times 100\,(\%)$$

For work done under contract this is a theoretically straightforward measure which is often applied in practice. The same measure can also be used for direct labour work but in this case care must be taken to distinguish between the value of work done and its cost (these may differ due to inefficiencies or price variations). Physical progress in this case cannot be equated with financial spending, since this ignores the possibility of cost over-runs. To measure physical progress as a value of work completed, each line agency may need to develop its own physical measures of progress, relevant to its sector, eg number of schools completed, number of health workers trained.

In the third situation, the output of the activity cannot be directly valued. This might occur, for instance, in supply only contracts or in projects such as the development of an extension service (in which the activities might include design of the programme, recruitment of agents and training). In this case the targets and progress should be expressed in terms of milestones marking the end of each activity or, if this is not possible, in terms of

$$\frac{\text{Time spent to date}}{\text{Total time to complete}} \times 100(\%)$$

The problem of using time as the basic measure of progress is that time spent to date on an activity may bear little relationship to actual progress towards completion of that activity. It is, for instance, only too easy to think of a situation where half of the scheduled time for an activity had been spent, but no meaningful progress towards its completion had been made. Using time as the measure, this would be recorded as 50 per cent complete, which would be wholly misleading to the manager concerned.

Whatever method of measuring physical progress is selected, project managers need to institute a simple system which allows comparison of the planned target against the actual target achieved (Table 9.2). This can be done on a monthly or quarterly basis, or any other time period which seems appropriate to the project. Such a pro forma can easily be computerized, but care must be taken to ensure that computerization does not detract from the essential simplicity and utility of the concept.

Table 9.2 Monitoring pro forma

Activities		Unit	Cumulative progress (by quarter)			
			1st	2nd	3rd	4th
Formation of production	Plan	No	—	2	3	5
co-operatives	Actual		—	2	2	
Contract for construction	Plan	%	30	70	100	100
of stores	actual		20	50	80	
Procurement of	Plan	*	A	B	C	D
milling equipment	Actual		A	B	B	

* Milestone code	A =	tenders advertised
	B =	contract awarded
	C =	equipment delivered
	D =	equipment installed and working

The chart shows monitoring data entered up to the end of the third quarter, when all three activities are behind schedule.

Measuring financial progress

As well as monitoring physical progress, managers will be concerned to measure financial progress, both to relate it to the total project budget and to ascertain the costs of individual items and activities within the project and how these compare to the original estimate. Once again, therefore, such a process will require the preparation of a financial plan – in other words, a budget – for the project, and subsequent measurement of actual expenditure and comparison against budget.

Once the budget has been prepared and the project is under way, the project manager will need a cost reporting system to provide information on actual costs. Unfortunately it is often not possible to rely directly on the accounting system to do this, for a number of reasons. First, such a system is likely to have been established prior to the project and to be used for many other activities of the sponsoring ministry or organization. It is thus unlikely to record costs for individual items in the way that the project manager would find useful. Second, accounting systems are often slow, so that costs are recorded and reported several months after they are actually incurred, too late to allow corrective action if variances are observed. Finally accounting systems do not record costs committed but not yet incurred. For instance, costs which are committed on the signing of a contract may be a main element in the overall build-up of costs. Project managers will therefore need to establish a cost reporting system which provides the information required (costs incurred and committed, for each item and activity in the work breakdown structure) in a timely fashion, say within a month. Public sector managers will need to take particular care that they keep a measure of overall spending on the project against overall budget. The nature of the annual public budgeting system in many developing countries means that managers are not able to spend above each year's authorized budget, so that cost over-runs are met by delaying implementation (thus passing part of the over-run to the next year), or reducing the scope of the project to compensate. In this type of system it is, therefore, easy to lose sight of the total cost of individual project components and activities, and indeed the project itself.

Monitoring project 'quality'

The last element of the project monitoring system concerns quality. Project managers will be concerned to ensure that the outputs provided by the project will be effective in producing the effects and impacts intended by them. The system required to ensure this will vary from project to project. In the case of physical construction there are established systems of supervision, testing and checking against the original specification, and a growing number of techniques of 'quality assurance'. In the case of projects with institutional outputs such as new delivery systems, trained personnel and the like, project managers will need to derive specific systems of spot checks and surveys in order to monitor satisfactory project quality. For these projects such monitoring is particularly important, since it provides the mechanism by which managers receive feedback on project performance and can judge whether the outputs being achieved are actually likely to contribute to the desired effects and impacts. This in turn may lead to the possible redefinition of project processes for adaptive and people-orientated projects.

The control system

The control system uses the output of the monitoring system to provide project managers with the basis on which they can actually manage their projects. Control is perhaps a rather unfortunate term to use for this process, since, in other contexts, it has connotations of restriction and repression. In the context of project management, however, it describes the process of analysing the variances shown up by the monitoring system, assessing the implication of these variances for project progress, and taking remedial action where necessary. In many ways it therefore lies at the heart of the project manager's job. In those cases where remedial action cannot be fully effective in restoring the project to its original plan, then managers will need to establish a revised plan (with, for instance, a delayed completion date or an increased total cost), against which future project progress can be monitored. It is worth stressing that behavioural skills are required of project managers in this process, as much as technical and modelling skills. Any revision in the baseline plan must be communicated to the project staff and accepted by them, a process of change which may well not be smooth.

As far as control of physical progress is concerned, the key data with which managers work are the revised estimates of time to completion, both for individual activities and the complete project. These should be calculated for each activity based on the progress for that activity: in many cases this may not be a simple calculation, but will require further data collection and judgement by the project manager and project staff.

The basis of the review is the project bar chart whose construction was described in Chapter 6. As monitoring information becomes available, each activity is shaded so that the unshaded portion represents the remaining time required to completion (Figure 9.2). Progress is then compared with a date marker moved along the time chart. Construction of a new network and time charts may be necessary if serious dislocation of project progress occurs. In this case the analysis is repeated with the remaining durations entered for each activity, instead of the original duration, and the 'time-now' is used as the start date for the analysis. Activities which have been completed have a remaining duration of zero.

The remaining time to completion is the key parameter in progress control. In a similar way the revised estimate of the cost to completion is the most important factor in project cost control. At all times, managers and owners need to know the total requirement of funds required to complete the investment: this remains true even when implementation is in progress and part of the funds have already been expended.

167

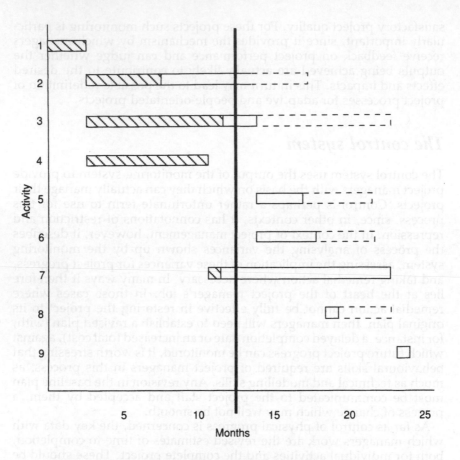

Activities completed or in progress are shaded so that the portion left unshaded represents the time required to complete that activity at the time of monitoring (month 13). These data show that all these activities currently in progress (2, 3 and 7) are behind schedule.

Fig. 9.2 Bar chart with monitoring data

It is unfortunately true that, for many development projects, financial progress to date does not provide any reliable data for use in estimating the remaining cost to completion. Frequently projects suffer such severe dislocation that managers are forced to make completely new estimates of the cost of the activities or part-activities remaining.

In conditions where good physical progress and cost data are available, and where the project implementation environment is sufficiently stable, it may, however, be possible to make useful estimates of costs to completion by using past data. This is done by means of variance analysis.

Performance variance distinguishes between the value of work planned to be completed and the value of work actually completed at the monitoring date. It can be expressed as an absolute value or a ratio. A negative variance (ratio less than unity) indicates a delay, which can be confirmed by the physical progress monitoring data: indeed it is those data which are used to calculate the performance variance. If indirect time-related costs are included in the project cost estimate, a delay signifies a cost increase equal to the additional time multiplied by the time-related costs. This increase is sometimes known as the 'cost of time over-run'.

Efficiency variance distinguishes between the value of work actually completed and the actual cost of completing that work. It answers the question 'are the various project tasks being executed within the originally estimated costs?' Again, the variance may be stated as an absolute sum, or as a ratio. A positive variance or a ratio greater than unity ('cost over-run ratio') indicates a cost increase. Such cost increases may be due to inflationary pressures, more difficult working conditions, changed methods and the like. It is worth noting that, for works carried out under contract, efficiency variances are borne by the contractor. The efficiency variance may be used, together with the revised programme to completion, as a basis of a revised estimate of the costs to completion. This calculation may often assume that no further delays will be incurred, that cost over-runs will continue at the same level as before for those activities already in progress and that no cost over-runs will take place on those activities already started. Of course, all these assumptions need verifying and indeed the analysis may be modified to take account of different assumptions.

When the analysis has been completed, the results may be presented graphically on the same figure as the original 'S' curve, distinguishing between the performance variance and the efficiency variance (Figure 9.3). The remaining cost to completion (budget) and time to completion (schedule) then form the basis for the project manager's revised project baseline, against which future progress should be monitored.

Project management information systems

Project management information systems (PMIS) can undoubtedly play a vital role in assisting managers to implement projects effectively and efficiently. Nevertheless there are some dangers inherent in the concept of PMIS, particularly related to the use of computers and information technology, and careful thought is needed if a useful and practical system is to be developed.

While modern computer technology is an extremely powerful tool, it must be borne in mind that its particular feature is its capacity to handle and process data, whereas the PMIS is concerned with information which is of value to the manager in monitoring and controlling the project. Figure 9.4 summarizes this distinction. Data are a record of a primary transaction such as 'height, weight and age' or 'date, payee and amount'. Such data have a cost in their collection, and do not, by themselves, provide information. Information, by contrast, involves the aggregation, analysis and presentation of data, and the transmission, receipt and interpretation of data by receivers in such a form that the data are of value

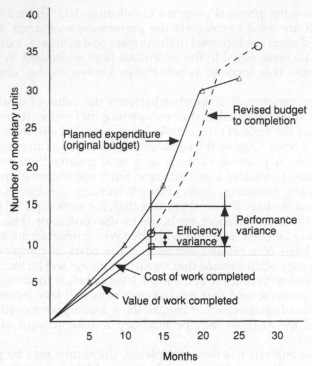

Fig. 9.3 Project cost control

to them in their particular situations. This distinction, which is often not fully appreciated by managers and system designers, can be summarized in such sayings as 'a surfeit of data but a lack of information' or 'what I want is information – all I get is reports'. Indeed very many so-called management information systems in existence in development projects today are, in fact, very detailed and comprehensive systems of record-keeping (costs incurred, activities completed, etc), rather than systems providing useful and forward-looking management information.

A good example of the problems that can occur when the distinction between data and information becomes vague was observed on the management of the Rice Canal Irrigation System in Pakistan. A management system has long been in operation there, designed to give the manager information on water flows throughout the system. During the five months of each year when the canal is in operation, readings are taken at the seventeen major water control structures on a two-hourly basis. Each set of readings requires three measurements (upstream level, downstream level and gate opening to allow the calculation of discharge). Annually, therefore, approximately 90,000 bits of data are read and transmitted back to the engineer's office, where they are systematically entered in a large book. Unfortunately, however, not one piece of data analysis or processing is carried out so that the data remain precisely that

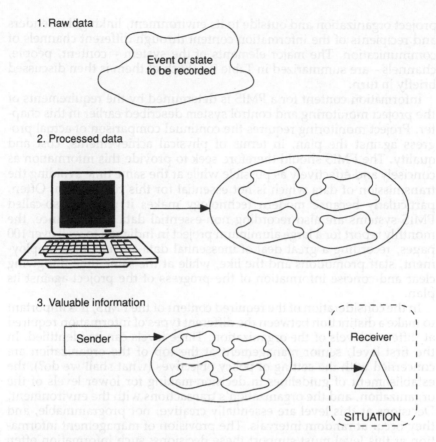

1. Raw data

Event or state
to be recorded

2. Processed data

3. Valuable information

Sender

Receiver

SITUATION

Fig. 9.4   Data and information

– data – and provide no information to help the engineer manage the canal system more efficiently.

A further reservation about the value of a computer-based PMIS concerns the necessity for keeping all available channels of information open, and not just those which process and transmit data collected on a routine and regular basis. This is often termed 'hard' information but much important information will become available through other means such as informal contacts and public media of communication (this is sometimes termed 'soft' information). Many managers appear to be successful to the extent that they are able to tune into and use soft information that comes to them through a variety of external and ad hoc sources, which provide them with insights into the constraints and relationships of importance to their projects.

In designing a PMIS it must be kept in mind that it is a system for storing, processing and transmitting data to provide information for managers in a way which will assist them in their decision-making. The PMIS therefore encompasses not only the data-handling mechanisms but also the channels of communication – up, down and sideways within the

171

project organization and outside to its environment, linking the senders and recipients of the information content through different channels of communication. The major elements of the system – content, people, channels – are summarized in Table 9.3. Each of them is then discussed briefly in turn.

Information content for a PMIS is determined by the requirements of the project monitoring and control system described earlier in this chapter. Project monitoring requires the continual comparison of actual progress against the plan, in terms of physical achievements, cost and quality. The PMIS should therefore seek to provide this information as concisely and effectively as possible while at the same time avoiding the transmission of data which is not essential for this comparison. Often, particularly because modern technology makes it possible, so-called PMIS systems are also recording non-essential data. For instance, the monthly report for a large aluminium project in India ran to well over 100 pages, recording a great deal of inessential data such as total employment, staff promotions and the like, while at the same time not giving clear and concise information of the progress of the project against its plan.

In the consideration of the required content of the PMIS, it is important to make a distinction between the different types of information required at different levels of the organization. Three levels can be identified. In the first level, senior management at the top of the organization are concerned with the setting of policy objectives (what shall we do?), the establishment of guidelines in decision-making for lower levels of the organization, and the organization's transactions with the environment. Decisions at this level are essentially creative, not programmable, and they occur at random intervals. The provision of management information at this level must support these decisions: such information often comes from without, rather than within, the organization, and may originate from many diverse sources.

The second level of information requirement is the project or middle management level. Here decisions are concerned with the establishment of procedures and allocation of resources in the pursuit of the overall project or programme goal and subsequent monitoring and control of progress towards that goal. Information requirements for this level may be in the form of summaries or aggregates, often highlighting exceptions which may require remedial managerial action. Much of the information will be generated from within the organization, though there is still a good deal of interaction with the external environment.

In the third level, the field and site staff are concerned with executing requirements through task work. Decisions are short term and programmable: the information to support them is generated from within the organization through established routines and procedures. The requirement is for detailed information on a regular basis; there is comparatively little interaction with the environment (one of the functions of the higher levels of management is to mediate the relationship between the site staff and the environment).

Table 9.3 Project management information systems

Level	Function	Content	Output	Frequency	Channels
Top management	Strategic planning for the organization	'Soft' information often from external sources	Policies Constraints	Irregular	Written and verbal reports Meetings
Project manager	Project management	Aggregated data from site Summaries Exceptions Milestone charts	Decisions Schedules Resource allocation and control	Monthly Quarterly	Written and verbal reports Meetings Site visits
Site supervision	Task execution	Task progress against plan	Actions Resource utilization	Daily Weekly	Written and verbal reports

173

The different managerial levels require information at different frequencies. First, top management, concerned with long-range policy-making and strategy, require information randomly and generally at long intervals. Second, project managers require regular reports at medium-term intervals, perhaps once per month. Third, field and site staff are involved with regular information at short intervals, weekly or even daily. Increasing the volume of information does not necessarily lead to better decisions, particularly at the senior or middle levels, where it may simply result in information overload. Increased information is normally more useful at lower levels where essentially repetitive decisions are made.

Channels or media of communications used by the PMIS are the same as those used by project managers in their day-to-day management communication, with the important addition of the possibilities offered by information technology. Managers will place heavy reliance on the various forms of verbal communication available, ranging from informal contact face-to-face or at a distance, to a regular and formal programme of meetings of visits. Many construction projects, for instance, have a formal system of monthly site meetings between the contractor and the owner or owner's representative to discuss progress. In other situations a manager or owner may institute a programme of site visits to provide a regular forum for discussion on progress, reviewing achievements measured against the plan and identifying future actions.

A PMIS is likely to give greater importance to written forms of communication, however, since these also provide a ready form of referencing and recording which may subsequently prove useful. Written communication ranges from memos and letters to reports and it is now customary to have monthly or quarterly reports recording progress on projects and providing monitoring data. Of course, modern information technology provides a great number of additional channels for passing 'hard copy' of information, including distance processing, electronic mail, facsimile systems and the like. Care must nevertheless be taken to ensure that the technology is not used to produce large quantities of data which are very time-consuming to read and which may tend to obscure the very important information which the managers need to know.

While the content, users and channels of communication are the basic elements of the PMIS, it is also necessary to consider the system in relation to the whole project organization. Indeed the type of PMIS which is established may well have a fundamental effect on the type and behaviour of the organization itself. In this consideration, two aspects are of importance. First, within the project organization itself, should the PMIS be set up and controlled through a separate part of the organization or directly through the functional departments? It is common to find, particularly in organizations which are concerned with the implementation of large projects or widely different programmes, that a separate department has been set up to take responsibility for project monitoring and control. This department measures progress in each of the main areas of the project and generally reports directly to the project manager. While there might seem some justification related to the need for specialist expertise to be grouped together in a single organizational location, the effect of this is to introduce an element of confrontation between the functional departments and the monitoring department. The former see the latter in an adversarial role because of their direct reporting relationship to the project manager, and may even try to conceal or misreport adverse variances, in case these may seem to reflect unfavourably on their own performance.

Public sector organizations implementing projects are, by their nature, bound to be somewhat bureaucratic and as such, are liable to sanction those who fail to perform their duties correctly, but do not, conversely, directly reward those who perform effectively. Concealment of failure may therefore be a powerful motivating influence on managers' behaviour if they are separated from the monitoring and reporting system. Ideally, by contrast, the PMIS should be of assistance to the managers themselves. As suggested at the beginning of this chapter, this will happen only if they feel an 'ownership' of the PMIS, in relation to its design, establishment and operation from within their own departments, so that they view it positively rather than as a threat.

A separate, but related, aspect of this discussion is the role that information technology can play, particularly in linking projects and programmes back to national monitoring systems. There is a continued interest in national monitoring systems, deriving from the initial work done by Malaysia with the Operations Rooms system. Increasingly, countries, particularly in south and south-east Asia, are attempting to set up systems which provide centralized monitoring information for all projects. Here again, though, there is the risk of the manager being separated from the PMIS. If all monitoring data are to be transmitted back to the central monitoring organization (CMO), it is a short but unfortunately logical step to suggest that relevant managerial action should also be taken at the CMO, rather than by the project manager at site. Indeed, with many highly bureaucratized systems, the tendency is for this to happen anyway, and the tendency may well be reinforced by the opportunities provided by information technology to pass information rapidly from the project to the CMO and vice versa.

It is worth bearing in mind, however, that information technology is relatively neutral in this respect and its effects are determined by the way it is used. At the present time, the trend is to see it as a tool for passing information from the projects to the centre, thus increasing centralized control. The technology can, of course, just as easily pass information from the CMO to the projects, thus providing decision-support information at the periphery and increasing delegation and decentralization.

Information technology in project management

Modern advances in information technology clearly have major implications for development projects in general and their management in particular. The potential of technology to store, process and transmit large quantities of data, the speed with which this can be done and the ease with which widely separated locations can be linked together make possible a degree of managerial involvement which was unthinkable as recently as the early 1980s. Nevertheless, as discussed earlier in the chapter, there are dangers inherent in the use of information technology, including its strength in handling data rather than providing information, and in particular its inability to provide 'soft' information. These dangers must be fully appreciated if its full potential is to be realized.

Consideration of the value of information technology in project management is conveniently broken down into a discussion of hardware (the physical equipment of computers and peripherals such as printers, fax machines, etc) and software (the programs that run the equipment). In both areas very rapid advances have taken place over the last few years which have revolutionized their applicability for managers. Looking first at the hardware, the main development has been through miniaturization. Whereas, in the past, any significant quantity of electronic data processing required a 'main frame' computer, which often filled a small room, nowadays these are required only for really large applications such as big administrative systems, and complex technological calculations. By contrast, individual project managers can normally have access to all the processing capability they need through 'personal computers' or micro-computers, which are only the size of typewriters and can be located at the manager's desk. The power of these machines is being increased still further through advances in 'networking', which links together numbers of micro-computers so that they can communicate with one another and share data. Moreover, these machines are much more robust than earlier generations of electronic data processing equipment. While they are somewhat sensitive to dust and should be covered when not in use, they no longer need to be located in air-conditioned surroundings, so they are relatively cheap to instal and can be made readily accessible to all potential users. They can also be moved when required, indeed many are designed to be fully portable. In addition they have few moving parts and are fairly easy to maintain, so that they are likely to be increasingly available to project managers, even at the most remote project sites.

The electronic revolution of the 1980s has perhaps been even more dramatic with respect to software, than it has been through miniaturization of hardware. Whereas it was once necessary to have a knowledge of programming to use electronic data processing, nowadays the project managers can buy a range of off-the-shelf programs which will perform most of the information processing which they require. Many of the standard packages cost less than $500. In the exceptional cases where additional manipulations are required, these can often be written to order by software firms.

The standard software programs which are of most value to project managers can be divided into five major categories: spreadsheets, databases, word-processors, graphics programs and project management programs.

Spreadsheets

These programs are the modern equivalent of calculators, providing extensive calculating power on a matrix of rows and columns. They are ideal for financial calculations of all sorts and other activities where numerical processing is required. The market leader in these programs is currently Lotus 1-2-3.

Databases

Databases, of which the best known examples are the dBASE programs manufactured by Ashton-Tate, provide for the storage and retrieval of data, together with a limited amount of processing capability. The data-holding capacity of these programs is very large, though they do not, by themselves, provide a PMIS since the data they hold have to be manipulated to provide information.

Word-processors

Word-processing programs are the modern equivalent of typewriters, providing for the preparation, presentation and storage of text.

Graphics programs

There are a wide range of graphics programs, often linked to word-processors and sometimes spreadsheets. These provide for graphical presentation of data in a variety of forms. Such facilities are clearly of increasing importance to managers, given the important element of their job which involves communication, particularly with the project environment.

Project management programs

There are a variety of programs on the market which provide for critical path analysis of projects. While many of these are fairly cheap, they are unfortunately of limited applicability for development projects. This is for two main reasons. First, they are extremely detailed, thus providing a level of managerial involvement which is not appropriate for managers of development projects (for example, they often assume that the durations of the various project activities are likely to be measured in days, whereas in many development projects, weeks or months would be more appropriate.) Second, they are written from the viewpoint of contractors, actually carrying out the work, rather than the owner's project manager, supervising and controlling it.

Other packages

While these are the main groups of programs likely to be of interest to project managers, mention should be made of the so-called 'integrated packages' which are a combination of spreadsheet, database and word-processor, perhaps with some graphics capability included. Finally there are also a number of statistical programs which are readily available, providing a range of statistical processes and tests. These, however, are not likely to be required very often by project managers, as they have more applicability in the analysis of comprehensive monitoring data from existing schemes and systems.

Applications

While itemizing the available software programs, it is also useful to think of the range of applications which are specific to project management, and where these packages might be useful. One obvious application might seem to be the project management programs which provide a programming and scheduling facility. However, for the reasons given above, the programs which are readily available on the market are not particularly useful. Spreadsheets, on the other hand, provide an invaluable tool, not only for many aspects of financial planning and management, but also for a variety of technical aspects. They can be used, for instance, to prepare cost estimates, which can be constantly and instantly updated as new information comes in. Spreadsheets can also be used for a variety of numerical modelling applications, where the effects or outcomes of different managerial decisions can be investigated. Databases provide a useful tool for record-keeping and for those aspects of project management which utilize large quantities of data, such as stores-keeping, stock and inventory control. Finally word-processors provide a major facility for project managers in their role as communicators. Indeed, it may very well be that a word-processing program is the first one to be installed on a project manager's computer.

Further reading

Bowden P 1988 *National monitoring and evaluation.* Avebury, Aldershot.
Casley D, K Kumar 1987 *Project monitoring and evaluation in agriculture.* Baltimore, World Bank.
Mintubise T 1984 *Managing information systems.* West Hartford, Connecticut, Kumarian.
Smith P 1984 *Agricultural project management: monitoring and control of project implementation.* London, Elsevier.
Tricker R I 1982 *Effective information management.* Oxford, Beaumont Executive Press.

CHAPTER 10

Managing people in project organizations

Farhad Analoui

Introduction

People are the basic ingredients of an organization. It is for people's sake and to their benefit that development projects are undertaken, and without whom very little can be achieved in so far as the implementation of these projects is concerned. Until recently followers and advocates of the principles of scientific management put emphasis on the knowledge of the 'task' rather than understanding the role of 'people at work'. When things went wrong, it was generally the manager's technical competence and not managerial skills which were open to question. More recently it has been recognized that people are not only subordinates, but also the essential resource available to managers for transforming ideas, inspirations, materials, capital and technical competence into products or services.

Hammond (1990), an organizational psychologist, asserts that if the obvious differences between projects, that is those to do with intrinsic difficulties such as the limitation of funds and technical skills, are eliminated, 'the most glaring factors accounting for why some projects are more successful than others are those to do with people'.

Often development project managers are recruited primarily for their technical expertise and not for their skills in managing people. Many assume that making the optimum use of people requires no special training or skills, yet experience indicates that most problems associated with the management of projects are primarily people-related. Managing people is, therefore, an essential managerial task. This is especially so in the project context where the situation changes and develops over short periods of time along with the acquisition and attitudes of people working in them. Project managers need to be prepared to take responsibility for managing dynamic situations as they interact with the human resources of the project. The project manager will have to deal with individuals both

179

within groups and teams, and outside the project organization, all having their own expectations from the manager and the project. It is, therefore, important for managers to understand why people behave in the way that they do. Why do some work hard while others produce only the minimum required? Why are some people almost unaffected by pay and other work-related conditions, and require only appreciation and recognition for what they do, where others seek power and status? How can people be motivated and led so that they give their best performance to the project? The answers to such questions require an understanding of human behaviour. People's performance and effectiveness within the organization cannot be expected just because they are hired, employed or otherwise associated with a project. To get the best from people a manager needs to communicate effectively, manage meetings, lead by exemplary performance and conduct, motivate, handle grievances, understand interpersonal needs, act assertively, solve problems and make decisions. In order to do all this managers need to interact successfully with others, either on an individual or collective basis, in the context of work and within the project environment.

The remainder of this chapter provides project managers with the basic knowledge and skills to enable them to deal effectively with people-related aspects of projects.

How do we expect people to behave?

McGregor (1961) has provided an interesting insight into the behaviour of people within work organizations. He suggests that, to a large extent, it is people's own assumptions about others which determine styles of managing them at work. Consequently, it is managers' beliefs concerning their subordinates that largely determine the nature of their responses towards them.

McGregor suggests that there are basically two sets of assumptions that managers make about the world of work and the people in particular. These are referred to as Theory X and Theory Y and have been briefly considered in Chapter 5. According to Theory X, people are believed to basically dislike work and to avoid accepting responsibility for it, if they can. Advocates of Theory X strongly believe that people need to be directed continuously and their behaviour monitored and controlled. Therefore, a manager with Theory X assumptions about people and their attitudes to work, will resort to either coercive or seductive styles for managing people. Punishment and reward – carrot and stick – are the basic tools for such managers.

Theory Y on the other hand assumes that people are naturally active and enjoy achieving goals and are committed to the objectives of the organization. Moreover, from this perspective, people are viewed as being creative and responsible. Theory Y type managers see their role as being responsible for ensuring that objectives are clearly defined and work is well co-ordinated. As far as possible, they will attempt to involve people in the process of decision making. The application of Theory Y has profound implications for managing people at work.

Whatever theory managers adopt there is a clear requirement for them to develop a capacity for self-diagnosis and the generation of self-awareness in order to determine what basic assumptions they have concerning people and their attitude to work.

Frames of reference

Frames of reference, a phrase originally coined by Goffman (1974), refer to an individual's repertoire of values, thoughts and beliefs. It is argued that frames of references are responsible for the perception of social reality, quality of interaction with others, interpretation of others' responses and provide the basis for people's actions at work. A large proportion of people's values are generated and influenced by environmental forces present both within and outside of the organization. Family, education, religion, political beliefs and ideologies, and most importantly cultural values, from tribal belief to that which is shared by a large section of society, are collectively responsible for making up people's frames of references. This provides the very basis for the ways that people perceive events and situations in which they act and interact with others. The workplace is one such situation.

It is people's unique perception of the world of work which determines the way in which they go about interacting with one another. The implication of this for managers is that they should possess sufficient knowledge and awareness concerning the dominant social values and beliefs which people hold. Understanding people's frames of references, attitude and orientation to work can provide practitioners with valuable clues for anticipating how they are likely to behave.

Managers and communication

It is argued above that project objectives cannot be realized unless managers know how to deal with people effectively. This requires the ability to persuade others to perform their functions effectively as and when they are needed. As this involves communication it might be argued that this is the most important function of management. The achievement of organizational goals largely depends on the ability of the managers to ensure that people are aware of, first, the objectives of the organization or project, second, the correct ways in which the optimum contribution can be made, and third, the extent of their contribution to the overall success of the enterprise.

Bad and ineffective communication results in the creation of barriers and misunderstanding between managers and the employees. Once barriers are established they tend to become more difficult to overcome as time passes. Managers should be able to avoid the creation of such barriers by utilizing the principles of effective communication.

181

Formal and informal systems

Communication can be divided into two types: formal and informal. The formal or prescribed pattern of communication works through the project hierarchy. The structure and design of the project organization, whether it is functional, matrix or projectized, determines the form and levels of communication in that organization.

Formal communication is used, often in written form, to approve decisions, to record the background, reason and conditions under which decisions are made, to disseminate information concerning the policies and procedures which affect the conduct and behaviour of people, such as the working patterns, conditions and objectives of the project.

At the project level a formal system of communication is necessary in order to prepare applications for funds, deal with payments, prepare reports, and provide background information for relevant committees. As a rule, the larger the size of the project or the administrative organization, the greater the requirement for a formal system of communication to manage work processes and people.

Informal communication is not merely an alternative form of communication, but a necessary means by which project managers handle human resources. With recent advances in the fields of micro-chip technology and telecommunications, managers do not have to rely only on face-to-face contact to communicate informally. Radio equipment and telephones have facilitated the informal exchange of information.

In situations of face-to-face communication, only one party (the sender) attempts to transfer a message, often coded by the use of words and jargon with which both parties are familiar, to the other party (be it an individual or a group). It is important that the manager communicates clearly and ensures that what is said is understood correctly. This means, a sender – the manager – needs to receive feedback from the individual or group involved to ensure that miscommunication has not occurred. Handy (1985) lists a number of factors that can easily lead to the emergence of misunderstanding and thus result in the occurrence of ineffective communication. They are chiefly perceptual bias, immediacy of means of communication, distance, jargon and technical terminology.

One of the most important communication skills for managers is that of 'effective listening'. Managers often ignore the need for effective listening, but psychologists suggest that people like to be listened to as much as they wish to be told. Listening not only strengthens commitment to the organization and the task in hand, but also provides the necessary link for the reception of feedback, thus improving the overall understanding of the situation.

While there are no hard and fast rules as to how much daily communication should take formal or informal forms, project managers should be able to strike a balance between the two, depending on the issues to be communicated, the circumstances in which the communication needs to take place, and the frequency and need for such communication.

It is vital to remember that communication in whatever form should be confined to 'adult to adult' level rather than 'parent to child'. The latter, which is often of a critical and negative nature, tends to create tension in the receiving party and initiate defensive responses. Parent–child communication, commonly known as 'talking down to people', generally leads to the receiver having less inclination to accept what was said. Managers have to be sensitive to the emotions and feelings which surround issues arising at work, but should endeavour to deal with them objectively, fairly and humanely.

Managing effective meetings

Meetings and committees are facts of life, especially for project managers and are often dealt with formally. Managing such events requires special attention to facilitate meaningful communication.

Ineffective meetings are generally characterized by appearing to have no real purpose; complicating instead of clarifying issues; lasting too long; providing an opportunity for social entertainment instead of an opportunity to solve task-related issues. An ineffective meeting is recognized by delay in decision-making and taking action.

Managing people in a meeting requires certain conditions and skills on the part of managers. First, managers should ensure that the purpose of the meeting is clear; a short agenda needs to be prepared and distributed in advance to give the participants a chance to add to the list. People should be reminded of the reason they are attending the meeting and what they are expected to contribute to the discussion.

Second, the agenda should be itemized and reasonable time allocated to each item on it. Project managers in particular should avoid the occurrence of the time over-run – a characteristic of meetings held in bureaucratic organizations. One of the roles which managers should play in a meeting is that of being a 'time-keeper'. They should also avoid wasteful contributions.

Third, attendance should be restricted to those individuals who are likely to be affected by the issues being discussed. In both projects and administrative organizations it has been observed that regular meetings are attended by almost the same individuals over and over again. This often leads to the emergence of issues and discussions which are not helpful to the achievement of the goals of the meeting and to its use as an arena for personal and political exchanges among the participants. The minutes of meetings should be made available to those who are likely to be affected by the discussions or decisions made, to avoid unnecessary attendance.

Fourth, relevant information should be made available to the participants. When relevant information is not made available to the participants speedy decision-making is hampered and the whole process is subjected to unnecessary delays. It is of the utmost importance that managers provide participants with all relevant information prior to their attendance so that there is sufficient time to digest the issues. Understanding these need not, therefore, take place during the meeting.

Fifth, people show commitment to the activities which are seen to have visible and identifiable outcomes. Meetings which are adjourned unnecessarily to another time or tend not to achieve their objectives will result in the demotivation of the participants involved and the loss of faith on their part in the process of decision-making as a whole. Managers need to ensure that the objectives of the meeting are reasonable and are achievable. As far as possible, major objectives need to be broken down into attainable targets for discussion.

Sixth, whenever decisions are made, the necessary actions for implementing those decisions need to be initiated. Project managers should be aware that the overall aim is that of paying attention not only to what is to be done but also to how decisions reached can be effectively implemented. Successful managers often ensure that individuals responsible for taking action are invited to meetings and their commitment towards the implementation of the decisions reached gained there and then.

Seventh, meetings should be structured so that people see their role as team members clearly and not as a collection of individuals. In this way, people's contribution to a shared cause is encouraged and appreciated.

Finally, effective communication on the part of managers during meetings, whether the meeting involves an individual or a large group, depends largely upon their ability to convince people that what they wish to convey is important and ought to be taken seriously. Successful communication, whether at a one-to-one interpersonal level or in a formal meeting, provides the foundation for more effective interaction in the future.

Interpersonal communication

For managers the art of effectively interacting with others is both necessary and inevitable. Interpersonal communication provides the manager with an opportunity to pass and receive information and to establish sound and healthy working relationships. Both functions take place simultaneously and the attitude of managers towards other people and their communication skills determines the degree of success which they will experience in any given social encounter. Values and beliefs tend to act as the foundation for the perception of situations and events involving interpersonal communication.

Channels of communication

Passing and receiving information is necessarily considered as a crucial aspect of managers' jobs and not as something which they do automatically. Research has indicated that unless managers are in the position of possessing sufficient and accurate information, they are unlikely to make sound decisions.

Whenever two individuals interact with one another, be it in the workplace or elsewhere, large amounts of information are exchanged. Yet most information available to each of the parties either goes unnoticed or is misunderstood.

Albrecht and Boshear (1974) argue that in all social interactions people use four separate types of communication: these are facts, feelings, values and opinions. All four are used in an interaction, although people may not be aware of them.

Facts are different from feelings in that the latter represent the emotional responses to the situation and, therefore, may vary from one individual to another depending on how they are approached, perceived and affected by them. Facts are of an objective nature concerning a given situation and 'are believed to be true'.

The main difference between values and opinions can best be described in terms of the degree of their permanency. Values are permanent, strong beliefs concerning what should or should not be done. Opinions are short-term beliefs which are related to specific situations and are likely to change as the amount of information available increases.

One of the problems faced by managers is the confusion between the different types – in short to misunderstand others. For example, opinions should not be taken as facts. The responsibility always rests upon the receiver to check the information for accuracy and reliability. People often include their feelings in communication when discussing objective issues at work, thus 'personalizing' the data in their possession. Personal values and beliefs can obscure the importance of work-related issues.

Use of terminology unfamiliar to the receiver can also cause problems leading to misunderstanding. It is important for managers as senders of information to ensure that facts are clearly communicated, whether in the form of verbal or written instructions, feelings are distinguished from values and that opinions are not interpreted as beliefs.

In addition to the four basic types of communication, attention must also be paid to listening skills, as suggested earlier. Psychologists suggest that the actual words used to convey a message may constitute as little as 25 per cent of what is said. Emotions and body language make up the rest.

If managers concentrate only on the spoken and written word they are likely to miss most of what people are really trying to communicate. Words reflect the intellectual side of a person and they help to create understanding between the sender and the receiver. Emotions, however, show how people feel about the issues to be communicated. In terms of relating to others, emotions act as a powerful means of making contact. People should be encouraged to indicate how they feel about work-related issues. Active listening is the art of not only paying attention to what is being said, but also of being aware of what is communicated in the form of emotions and displayed by appropriate responses in form of body movement. Bolton (1979) asserts that nodding the head, saying hm and smiling when listening to others encourages people to continue conversation. Bolton suggests that asking questions which are open rather than closed, showing approval of what is being said and generally being attentive to what others say will motivate people to overcome the problems of ineffective communication. Actions and behaviours provide the necessary visual impact and are used as a means of supplementing words and emotions.

It is not uncommon for people to experience difficulty in putting their ideas into words. Managers need to listen to all that is said and offer encouragement by providing the appropriate responses, that is verbal, emotional or behavioural, in order to make the process of communication and understanding as complete as possible.

Interpersonal needs

Managing people requires some understanding of what they do and why, especially in situations of work-related social interaction. Familiar examples include situations in which an individual has displayed the need to be in charge, or where an individual has shown a preference for being left alone. Some people feel more in need of being in control and others feel less in need of belonging to a group than others. But why? Bolton (1979) suggests that one of the skills which is necessary for the effective management of people is that of understanding their interpersonal needs. He asserts that three sets of interpersonal needs are paramount and most influential in terms of shaping behaviour towards others. These are the needs for 'inclusion', 'control', and 'affection'.

To satisfy interpersonal needs an individual needs to be with others, in a group, as a member of a team, in an organization and in the community. Indeed,`no man can be an island'. People are social creatures and naturally enjoy the company of others.

They generally wish to be included in the social system of the organization and naturally the extent to which an individual feels the need to be with others and to be a part of a particular scheme varies according to personality. Some people feel the need for inclusion more than others. An individual with a greater need for inclusion would perform much better as a member of a group, team or a 'task force'. At the other extreme, there are those who wish`to be left alone' and perhaps feel that, away from others, they are more likely to be productive. Such individuals would probably remain effective even if they had to work by themselves for a long period of time.

The above process is also applicable to the need for control and affection. In social interaction some people feel that they ought to have more control over what goes on, whereas there are those who prefer to be led and told what needs to be done. It is important to ensure that those who are placed in supervisory roles receive greater satisfaction from being with others and being responsible for leading and supervising the activities of others.

The assertive manager

Many managers indicate that at one time or another they have experienced the presence of what might be termed the 'after effect phenomenon'. This is usually experienced as a result of unsuccessful interactions. These are situations in which individuals feel that they had not been able to say what they really wanted to say, or did not do something even if they felt strongly that it ought to be done. Managers who experience such feelings report that it was the feeling of guilt and the fear of being classed as awkward and uncooperative which made them agree to a request even if they were sure that it was unreasonable. They assumed that good management of people is about agreeing with them and avoiding hurting their feelings. Others indicated a reluctance to say 'no' to their superiors.

It is interesting to note that most managers suffering from the 'after effect phenomenon' also agree that they had more jobs to do than time to do them. They suffer from excessive anxiety, work-related stress symptoms, feelings of low self-esteem and lack of control.

Socio-psychologists believe that such managers feel as they do because they are not being assertive enough. This group of managers tend to display submissive behaviour, which is an open invitation to others to violate their rights.

Other managers express the feeling that they do not enjoy the trust and full co-operation of their colleagues and their employees. They feel that people tend to avoid contact as though something is wrong with them. The explanation is that such managers have adopted what is known as 'aggressive behaviour' which is characterized by the violation of others' rights. A common reason given for such behaviour is that 'they had to make the point'.

Aggressive and submissive behaviour

Submissive and aggressive behaviour constitute the two extremes of the same continuum. These extremes do not generally guarantee successful results and will probably prohibit the formation of prolonged and sustainable work relationships. While the use of aggressive behaviour may result in immediate and short-term gain, it often leads to a deterioration in the quality of work relationships and the loss of confidence in the aggressor. Submissive behaviour, on the other hand, is a declaration of forfeiting rights and does not lead to strengthened work relationships.

Assertive behaviour

Assertive behaviour is the most satisfactory, least damaging and most productive form of interaction with others. Turner (1983) defines assertiveness as

the capacity to express our ideas, opinions or feelings openly and directly without putting down ourselves or others. It is standing up for your own rights in such a way that you do not violate other people's rights. It involves expressing thoughts, feelings and beliefs in direct, honest and appropriate ways.

Being assertive does not necessarily mean that people will never again be manipulated or manipulate others. Skills of assertiveness cannot be mastered instantly. They require constant awareness of feelings and needs and what others say or do. Often what drives people into becoming aggressive or to adopting a submissive stand is their inability to communicate skilfully their feelings, thoughts and ideas to others. Being aware of basic rights and what people should do to gain the respect of others by not violating their physical, socio-psychological and role-related authorities is the key to becoming assertive. Turner (1983) lists a number of what he terms as 'rights to be assertive'. These rights in fact refer to the ability of individuals to do things and behave in certain ways in order to protect themselves from being unnecessarily and unintentionally manipulated by others. He argues that assertiveness requires the ability to

1. *Be your own judge.* Being able to make decisions and adopt a responsible stand in relation to one's work, colleagues and the clients of the organization. This 'does not mean that we do not listen to others', but it does mean that people should not allow others to exercise unreasonable measures of control over their activities.
2. *Have the right to make mistakes* Being assertive is basically about being realistic and accepting people as they really are. Making mistakes is not an undesirable characteristic. Everyone makes mistakes and needs to face up and learn from them. People should not allow themselves to become victims of emotional blackmail because of mistakes made in their work. Admitting to mistakes and learning from them works better in the long run than simply trying to blame someone else.
3. *Be what you are* Often people are expected to be different just because it suits them. Assertive managers are aware of their capabilities, positive attributes and skills, as well as having a good knowledge of their shortcomings and weaknesses. It is important to create a positive 'winner' self-image which reinforces confidence. Managers who feel that they have to justify all that they do are not being assertive. Such an approach conveys a negative signal which subconsciously creates doubt in others about their capabilities. Such managers will then be in danger of being manipulated or being taken for granted.
4. *Be honest* Everyone is expected to make mistakes occasionally. It is also everyone's right to say 'I don't know' or 'I don't understand'. No one can know everything and indeed people are more respected for their specialized knowledge than their ability to make comments on all subjects and issues regardless of their proven knowledge. It must be remembered that if others are not successful in making themselves understood, it is unassertive to accept what they propose unopposed. Remember 'I don't understand this...' does not mean a lack of general understanding, but the inability and lack of skills on the part of others to make themselves understood.
5. *Other people's problem* Managers cannot be responsible for everything. They often face situations where they are confronted with an implicit expectation from others that they will solve their problems. Being an effective manager requires the ability to solve

problems, but it does not mean that managers should assume the role of 'resident trouble shooter'. Those who fail to recognize this simple point often find themselves spending precious time and energy solving problems for others. It is generally preferable to guide problem holders to where they may find the solution. On the other hand, psychologists explain how the excessive need for recognition from peers, colleagues and friends can make people vulnerable to their exploitation, albeit in the role of professional problem-solver.

6. *Say what you believe in* Being assertive is about respecting the opinions of others. Managers' opinions and beliefs are important and everyone should be entitled to give his or her opinion without having to justify it. Opinions must be respected by others, because they count.

7. *Be taken seriously* People expect others to listen to them and respect them for their knowledge and the contribution they make to the project's effectiveness and well being.

Becoming assertive puts people in the position of not being taken advantage of or taken for granted. It is often suggested that the effective management of people is about pleasing others and therefore acting submissively towards them. This is a myth. Managing people at work requires the knowledge of interpersonal skills, as a whole, of which assertiveness in particular is an important aspect.

Politics

In most work organizations, because resources are finite and people tend to pursue their own as well as the organization's objectives, it is only natural to discover that people are involved in competition to gain and maintain as much control as possible over the scarce resources available. People's interactions with one another determine the degree of control which is exercised over what goes on in the organization.

Politics, often referred to with negative connotations, require the ability on the part of the individual to adopt the appropriate response for a particular situation and deal with people and work situations selectively and decisively. Being assertive may mean consciously and intentionally surrendering one's right, at least temporarily. Should this action lead to gaining ground in the long term, it is justifiable. So, if needs be, managers have to say 'yes' when 'no' would be the most logical response.

Managing conflict

Managing conflict at work constitutes an inseparable aspect of any manager's job. All managers, even those with limited work experience, can describe situations in which they had to deal with discontented colleagues, subordinates, clients and even superiors.

After dealing with a conflict in an ineffective manner, managers experience mixed feelings of regret and remorse, for example, 'if only I had known how to handle the situation, I wouldn't have to be ...'. Surprisingly enough, managers encounter similar conflict situations over and over again, and they have to face the inevitable consequences of mishandling problematic situations if they are not prepared to deal with them effectively.

Writers on organizations and management have for decades been preoccupied with assisting managers to avoid such situations. The most basic questions which have been posed include what is the nature of workplace conflict? What are the symptoms of that conflict in an organization? What forms does discontent take, if any, and why? The most important of all is how can managers effectively deal with conflict at work?

The classical approach to managing people at work perceived conflict as an undesirable and destructive organizational phenomenon which had to be eliminated. Managers of this school deal with conflict situations by suppressing the manifestations of conflict through using their authority and pretending that it will not occur again.

Suppressing conflict

The application of classical management principles to managing conflict at work results in the virtual disappearance of visible discontent. But the unresolved conflicts and issues remain present and active, invisible to untrained managers, deep rooted within the very fibre of the work relationships.

While organizational conflict appears dealt with and so-called 'trouble-makers' have been handled accordingly, the organization often continues to become increasingly inefficient. People within the organization gradually become less effective and no longer show concern for the well-being of the enterprise or the achievement of its objectives. The era of early industrialization is fraught with incidences of workplace conflict which were unprofessionally managed and handled. In some cases the mismanagement of conflict even lead to the loss of lives and the livelihood of many people.

Organizational conflict and management of people

An important milestone in the development of the art of managing conflict took place when Mayo (1945) discovered, albeit by default, that among other things, the effective management of work organizations requires attention not only to organizational goals and objectives, but also to those of the people who participate in its activities.

This discovery paved the way for theorists, writers and management developers to view the organization and work relationships from a new perspective. Organizations were viewed in a more realistic fashion. They were seen as arenas where the constant interactions and exchange of views among the participants and their pursuit of their own as well as the organization's interests, had to be recognized as an inevitable aspect of organizational life (Katz and Kahn 1966).

From this new perspective, conflict at work did not have to be regarded as an alien and destructive force. On the contrary, the followers of the Human Relations philosophy viewed conflict as constructive, if only managers were able to recognize the legitimacy of its existence and accept the inevitability of its occurrence (see Chapter 5). More importantly, it was proposed that managers require the ability and knowledge to manage this organizational phenomenon skilfully through negotiation rather than with the liberal use of their managerial prerogative and authority.

Pluralistic frame of reference

Fox (1966) offered a comprehensive framework for understanding conflict and the attitudes of managers towards it. He compared the 'pluralistic' view, which derived from the work of the human relation schools, with the 'unitary' frame of reference which was adhered to by the followers of classical management. He then concluded that the rapid increase in the occurrence of industrial conflict, in particular strikes, can be explained by the failure of the management to accept the existence of sectional interests among the workforce.

The implications of this view for the management of people are manifold. Managers are expected to communicate with employees more effectively, create a climate of trust and be prepared to listen to, and to take notice of, the points of view of others, even though such views are often in complete contrast to their own. Also, managers need to be willing to negotiate the emergent issues with the representatives of the employees and their organizations in order to find an optimum solution to the organizations' problems. This was another indication of the need on the part of managers for skills of effectively managing people and tasks in the workplace.

Conflict and its probable causes

The most practical explanation for the emergence of conflict is that as people are unalike, they naturally hold different views from one another. Clashes of interest between individuals, groups, employees or employers constitute the basis for the occurrence of organizational conflict. Although conflict originates from various sources, it is the organization itself and work-related processes which account for the generation of most discontent among employees and employers at work.

Handy (1985) argues that the differences in the way people perceive and understand the objectives of the organization and the ways in which these stated goals are realized is one of the major causes of disagreement among people.

The achievement of one set of organizational objectives by one group of people often clashes with the interests of others. For example, increased effectiveness on the part of a production group may require more effort and an increased effectiveness from the suppliers of the raw materials, the maintenance department and perhaps the sales department. Another example of this situation is where the introduction of a new and more efficient method of realizing certain objectives will have implications such as redundancy and a loss of livelihood for some employees. In all, it must be remembered that since organizations' resources are limited, unavoidably some people will receive a smaller share than expected – a classic case for emergence of conflict.

Most discontent at work is concerned with the following:

1. diversity of the individual groups' and departments' objectives
2. diversity of roles (jobs) within an organization.
3. unclear contractual arrangements
4. duplicity of roles for individuals and departments
5. hidden agenda (undeclared objectives).

While these are the main causes for the creation of discontent among people at work, organizational analysts suggest that the most important and neglected cause of conflict is that of unclear policies and procedures concerning the attainment of objectives in the organization.

In most conflict situations, especially those which concern management and employees, it is mainly the lack of mutual understanding concerning what the organization stands for which creates the most disagreement and discontent. Management often takes it for granted that the policies are clear to all employees and that they are expected to work towards their attainment.

Territorial imperatives

Ardrey (1967) equates people's effort to acquire, expand and maintain control over the organization's limited resources, with that of gaining territorial rights in the animal kingdom. He asserts how conflict can be explained in terms of the violation of an individual's or group's physical (office), social (membership) and psychological (status) territory, by another individual or group. Andrey believes that this will result in the emergence of conflict amongst the people involved. By adopting the territory as a metaphor, he explains how ownership of territory is established by either 'deed', or 'precedent' or by both. Andrey identifies two important factors – overcrowding and territorial jealousy – as the main culprits of creating situations of discontent. In the organizational context, unclear job specification and the ambiguous nature of roles tend to create friction and discontent in the same way as can be observed in the animal kingdom.

It has also been observed that people in organizations tend to attach different values and importance to their jobs. The concept of 'dissonance', predominantly a psychological one, explains how an individual may begin to think that the grass on the other side of the fence seems greener'. There are many examples of this commonly experienced phenomenon in an organization where someone wishes to have someone else's job (territory).

Individual differences account for not only the differences in perception, but also preferences and the personal freedom to exercise a desired choice of action. These all add to the possibility that, at any one time, there is a strong likelihood that managers have to deal with conflict situations for which they may or may not be prepared.

How do managers approach conflict?

Conflict situations are defined as those in which the 'interests' of two people appear to be incompatible. In such situations the behaviour of individuals could be described as being influenced by two determinants:

1. *assertiveness* the extent to which individuals attempt to satisfy their own concerns
2. *co-operativeness* the extent to which individuals attempt to satisfy the concerns of other people.

Kilmann (1975) uses the above determinants of the behaviour to define five modes of dealing with conflict:

1. *competing* being assertive but uncooperative
2. *accommodating:* being unassertive but co-operative
3. *avoiding* being unassertive and uncooperative
4. *collaborating* being both assertive and co-operative
5. *compromising* being intermediate between both assertiveness and co-operativeness.

Everyone is potentially capable of using all five of the identified conflict-handling modes. Equally, no one can be characterized as having preference for only one single, rigid style of dealing with conflict. Managers often wish to know what is the best style of dealing with conflict situations and often find it hard to accept that there is no one best way of dealing with situations of discontent at work.

Each style has its own advantages and disadvantages. Basically, the styles which individuals adopt in a given situation result both from their personal predispositions and the requirements of that specific situation. Managers have been observed to shy away from confrontation in one situation as the best approach to avoiding escalation of a damaging dispute, whereas in another situation, the same managers were seen to stand their ground and confront the individuals or groups with whom they were in conflict.

Resolving conflict at work

Not surprisingly, managers are entrusted with the responsibility of dealing with conflict. The differences in individuals' values, beliefs and thoughts, their perceptions of situations and what they entail, the differences in the styles in which they approach the conflict situation, the differences in the nature of disputes, and most importantly of all, the unpredictable outcome of their interaction with one another, all make it extremely difficult, if not impossible, to put forward one single effective formula for dealing with all conflict situations. Conflict can be managed, but not eradicated.

Managers need not solely concern themselves with handling the most visible aspects of discontent at work, but they should strive to understand people and the nature of their interactions. In their capacity as managers they ought to create an environment which is conducive to co-operation and mutual acceptance of different and even opposing points of view.

Since managing conflict situations resemble dealing with problem-solving situations and necessitate the adoption of a rational approach towards them, managers may find the following principles helpful for effectively handling conflict at work.

1.   Identify the sources of conflict, people involved and issues con-cerned; these could be personal, organizational or ideologically based.
2.   Devise appropriate strategies that may be acceptable to all the parties involved.
3.   Create an ethos of co-operation and trust in which the differences of concerns and interests are heard and dealt with openly and frankly.
4.   Implement the strategies worked out.

Dealing with conflict is part and parcel of managing people in an orga-nization. Conflict results from differences in perceptions and can provide a constructive element in so far as change, innovation and creativity are concerned. If managed and channelled effectively, conflict will serve as a source of energy for achieving individual and organizational develop-ment and progress. If, however, the discontent is suppressed and man-agerial prerogative is used to condemn it, the conflict will become submerged deep in the work relationships and will inevitably gain beha-vioural expression which is of a covert and even unconventional nature.

Managers and motivation

Motivation at work concerns questions such as – 'What makes people do what they do at work?' and 'How and by what means can a manager help people to improve their work performance?' The term 'motivation' is usually loosely defined and is often used by managers as meaning to provide an incentive to get the job done. This is, however, a misconcep-tion of the motivational theories and their importance in terms of manag-ing people effectively at work.

It may be necessary to bear in mind that there is a direct relationship between the development of management perspectives, on the one hand, and managers' attitudes to work, people, and the ways which have been considered as viable and effective when managing people at work, on the other.

The advocators of the classical theories and scientific principles of management believed in using coercion or seduction as the two main means of ensuring that people performed well in their work. Financial rewards have always been considered as important motivators for achieving higher standards of work. The use of seductive measures, be it bonuses, employee subsidies, handsome pension schemes and the like, are all important, but as Hertzberg (1968) aptly suggests, they may not make people give their best, but only a fair performance.

It has long been recognized that the use of coercive measures such as the threat of dismissal, deduction of wages or salaries, and physical or verbal abuse, may result in a short-term growth in productivity on the part of the employees, but most certainly does not guarantee long-term commitment necessary for the accomplishment of the task. Such measures are likely to lead to alienation of employees and low morale among the staff.

Indeed, the question of what best motivates people in their work has more serious implications for project managers than it has for their counterparts in administrative posts, and routine production or process organizations. Project managers, especially those involved in development programmes, have to manage people in the face of scarcity of financial resources and limited time made available to them. Effective motivational strategies need to be used to ensure that a high level of performance is achieved when it is most needed.

Managers and project managers have often inquired, if money and punishment cannot ensure the achievement of improved performance, then what can? This is indicative of a further misinterpretation of the theories of motivation on the part of some managers, that is that financial rewards have no value as motivators. This is not so. Indeed, financial rewards can be used to ensure an initial improved performance; the drawback is that it requires frequent periodic reinforcement if it is to work over a long period. Each time more money is needed to achieve the previous level of performance, let alone a new, higher standard.

The use of punishment, too, tends to result in some form of 'diminishing return syndrome'. It must be remembered that adequate administration, effective supervision, fair performance-related rewards, healthy interpersonal relationships at work and the presence of good physical working conditions can all go towards ensuring the continued presence of acceptable levels of work performance.

In other words, as Hertzberg asserts, 'the use of hygiene factors such as money, simply keep people satisfied and they do not create unnecessary pain for them at work'. The hygiene factors (also known as 'maintenance factors') prevent people from feeling dissatisfied, but do not provide the necessary motivation for people to produce more than average. It is, therefore, argued that while every effort must be made by the managers to make the above conditions available, it is of the utmost importance that managers provide people with the ability and opportunity to experience some kind of achievement in their work.

Organizational psychologists state that people perform their best when they feel that they have achieved something at work. If the nature of the

195

work is not meaningful to the individuals concerned, they soon lose interest in their job and begin to experience alienation and job dissatisfaction. Thus, the nature of the work itself is one of the most powerful motivators which is available to the manager. In this respect project managers benefit from the fact that projects often create a stimulating and demanding environment in which to work. People in work organizations should be matched to their work. There has been the concern among theorists to match employees to their jobs, but recent research has revealed that managers, too, ought to be matched to their jobs and their organizations.

Analoui (1990; 1991a), in the research concerning senior project managers discussed in Chapter 5, discovered that the mismatch between managers and their organizations is clearly a contributing factor to the presence of low morale, ineffective managerial performance and the inability to achieve the laid-down targets. To achieve a desirable 'fit' between people and their job requires an awareness on the part of executives that the right person should be recruited to fill the right position. The provision of the right kind of training can often remedy a marginal mismatch between the individual and the job characteristics.

Responsibility at work is also referred to as a powerful motivator in the armoury of managers, but unfortunately managers have been observed to provide people with work-related responsibility without ensuring that their subordinates are capable of carrying out what is expected of them. People are intrinsically responsible beings and should be provided with stimulating jobs. Responsible jobs are ones which stretch people, but do not destroy their confidence.

Additional responsibility often acts as a challenge which motivates individuals to reach beyond their present abilities and standard of performance, thus bringing to the job the creativity and innovation needed for the future development of the organization. Of course, an experienced manager knows only too well that although the above motivators work well and will undoubtedly result in improved performance, they also bring with them another set of expectations on the part of the employees – the desire for advancement.

Promotion and advancement may not be an important issue in so far as managing people in a short-term project is concerned, but in a long-term association with an organization, people ought to be assured that they can develop and advance in their career. When people do not see any further opportunity for advancement in their jobs they either begin to feel trapped and look for an alternative job with attractive future prospects or they simply become ineffective in their present work.

Motivation does not solve the problem of people's relationship with the environment. It is about their aspirations, expectations, preferences and desire for ultimate development. When managers expect subordinates to work well and give their best to the organization, they should be aware that subordinates also need to be equipped and enabled to meet the expectations required from them in their work.

Although each work situation is different and people clearly have to be approached and treated differently, the following points should be remembered.

First, managers ought to diagnose the needs, abilities and preferences of their employees and attempt to match people to jobs which can intrinsically satisfy their work-related needs.

Second, people's performance and behaviour should be continually monitored and improved through training, promotion, job redesign and enrichment so that they can maintain a high standard of performance.

Third, the reward system in the organization needs to be applied fairly and, as far as possible, should be performance-related. Money and other tangible incentives are not the only source of motivation. Recognition of an employee's good performance, allocation of responsibility and advancement whenever appropriate all act as motivators.

Fourth, effective communication, at individual and organizational levels, must be ensured in order to understand the individual's needs, preferences and changes in attitudes or behaviours. Communication provides a vital link between the individual managers and the achievement of goals and objectives of the organization. Effective communication also ensures that people in their work feel that they belong, are appreciated and do matter.

Fifth, people should be provided with opportunities to expand their abilities and accept further responsibilities in their jobs. The degree to which the allocation of responsibility and exposure to challenging jobs can act as motivators largely depends on whether or not the individual's abilities and potential are well recognized, understood and assessed.

Sixth, people should be provided with adequate feedback concerning their performance, achievement and how they can become better performers. Effective communication is the prerequisite to the above.

Seventh, as far as possible people should be encouraged to achieve results because they enjoy doing so, and not solely because of the incentives offered. When people like what they do, they are intrinsically motivated, a point which is often grossly underestimated.

Finally, managing people, among other things, requires motivating and increasing their association with work and their commitment towards achieving the objectives of the organization. This requires understanding of their needs, abilities, potential and expectations which in turn creates the task-orientated behaviour. It is the task-related behaviour which achieves results and consequently the attainment of the goals of the project or organization, as a whole.

The manager as leader

As leaders of the organization, managers have to come to terms with the fact that people are needed to ensure the achievement of the organization's goals and indeed its development. The art of managing people is comprised of a bundle of abilities, skills and knowledge with which managers are not born. Successful managers explain how the process of becoming a manager inherently implies the responsibility for meeting people's present needs as well as ensuring their future development.

Can managers effectively deal with people by adopting an extreme classical position, 'tell' and the use of indisputable authority as the only means necessary for getting things done? Experience, observation and decades of research and development in the field of management proves that they cannot.

There is a range of styles and approaches available to today's manager for managing work through people. Tannenbaum and Schmidt (1973) explain how besides 'telling' style, leaders can also 'sell', 'test', 'consult' and even 'join' the employees, should they wish to seek alternative options, for getting the job done. Indeed, managers may decide to share the process of decision-making, consult with and invite subordinates' contributions for arriving at a final solution or utilize the ultimate extreme choice of complete participation on their part.

The concern for 'task' and 'people' as shown by Blake and Mouton (1978) can be combined to form the 'team management style' characterized by maximum concern for task and people, at one extreme, and the 'impoverished style' of management, characterized by a lack of concern for people and task at the other extreme of the continuum of leadership styles.

Managers, like any other individual within the work environment, have to make choices when confronted with situations which demand that they make a decision. From among management skills, managing people is probably the most crucial as it requires not only skills and knowledge, but also the ability to decide upon the most appropriate and efficient course of action. As Vroom and Yetton (1973) assert, managers ought to decide on the degree of involvement of the subordinate in the process of decision making. Thus, managing people also requires choosing from among the styles available to managers which in turn determines not only the degree of their participation but also the inevitable display of the leader's concern for people.

Does the effective management of people, therefore, require the manager to become totally involved with people-related issues at work? Or does such a tendency arguably result in the non-achievement of the task? Managers have always been interested in finding the right solution for becoming effective, yet the choice of becoming effective or remaining ineffective ultimately rests in their own hands. One useful piece of advice, however, is provided by the renowned management theorists Hersey and Blanchard (1982) who succinctly assert that ideal managers can modify their leadership styles to fit the needs of the situations. The adopted styles for the management of task and people should therefore be contingent upon the demands, constraints and choices which are realistically available to managers.

Conclusion

Only a learning manager can ensure that people in the organization are managed in such a way that they voluntarily and willingly contribute all they can towards the development of both themselves and the organization, as a whole. Managers are not born with the attributes, knowledge and skills necessary for managing people at work. Such qualities have to be acquired on an ongoing basis.

Whenever a problem is solved, a message is effectively communicated, a conflict situation is assertively defused and employees are properly motivated in their work, learning has taken place. Managers with the responsibility for their own learning, as well as that of others, as one of their major preoccupations, will consciously strive to develop their potential, abilities and skills to manage people in the context of work situations in an ever more effective and efficient manner.

References and further reading

Albrecht K G, W C Boshear 1974 *Understanding people: models and concepts.* La Jolla, California, University Associate Inc.

Analoui F 1990 Managerial skills for senior managers. *International Journal of Public Sector Administrators* **3** (2).

Analoui F 1991a Towards achieving the optimum fit between managers and project organisations. *Project Appraisal* **6** (4): 217–22.

Analoui F 1991b Making management training and development more effective: towards an interventionist approach. *Journal of Institutional Development* **11** (2): 217–22.

Analoui F 1992 *Industrial conflict and its varied expressions: a behavioural classification.* New Series Discussion Papers no. 21, DPPC, University of Bradford.

Analoui F, A Kakabadse 1991 *Sabotage: how to recognise and manage employee defiance.* London, Mercury.

Ardrey R 1967 *The territorial imperative.* New York, Collins/Atheneum Press.

Blake R R, J S Mouton 1978 *The new managerial grid.* Houston, Texas, Gulf Publishing Company.

Bolton R 1979 *People skills.* Englewood Cliffs, New Jersey, Prentice-Hall.

Child J 1984 *Organisations: a guide to problems and practice,* 2nd edn. New York, Harper & Row.

Fox A 1966 *Industrial sociology and industrial relations.* Research Paper no. 3, Royal Commission on Trade Unions and Employers Association, London, HMSO.

Goffman L 1974 *Frames analysis: an essay on the organisation of experience.* Harmondsworth, Penguin.

Hammond J 1990 The human side of project management. *Project Management Today* (IV) November-December: 12-14.

Handy C 1978 *Gods of management.* London, Pan.

Handy C 1985 *Understanding organisations.* Harmondsworth, Penguin.

Hersey P, K Blanchard 1982 *The management of organisational behaviour*, 4th edn. Englewood Cliffs, New Jersey, Prentice-Hall.

Hertzberg F 1968 One more time: how do you motivate employees? *Harvard Business Review* **46**: 53-62.

Katz D, R L Kahn 1966 *The social psychology of organisations*. New York, Wiley.

Kilmann T 1975 Conflict made instrument. In M Dunnette (ed.) *The handbook of industrial and organisational psychology*. Chicago, Rand McNally.

Lawler E 1978 *Motivation and work organisations*. Monterey, California, Brookes Cole Free Press.

McGregor D 1961 *The human side of the enterprise*. New York, McGraw-Hill.

Mayo E 1945 *The social problems of an industrial civilisation*. Cambridge, Mass., Harvard University Press.

Peters T, R Waterman 1982 *In search of excellence*. New York, Harper & Row.

Stewart R 1967 *Managers and their jobs*. New York, Macmillan.

Stewart R 1982 *Choice for managers: a guide to managerial work and behaviour*. New York, McGraw-Hill.

Tannenbaum R, W H Schmidt 1973 How to choose a leadership pattern. *Harvard Business Review* **51** (May/June): 162-80.

Turner C 1983 *Developing interpersonal skills*. Bristol Further Education Staff College.

Vroom V H, P W Yetton 1973 *Leadership and decision making*. Pittsburgh, University of Pittsburgh Press.

CHAPTER 11

Beyond projects: the wider context of management

Introduction

Throughout this book attention has been concentrated on the needs and skills of project management. The precise definition of a project given in Chapter 1 has led to particular emphasis on the management of project implementation or establishment, as the important phase of the project development cycle when capital is invested to create assets, facilities or systems.

Many of the problems and failures that have been encountered in project development stem from faulty management of the process of implementation, and it is this fact that has led to the emphasis in this book. Nevertheless, all managers must be aware that project development is not an end in itself, but a means to an end. The project cycle discussed in Chapter 1 can be a useful concept in helping planners and managers alike to see beyond the project phase and to appreciate that the end result of the project is the outputs which create assets of continuing benefit and value to the clients, and beneficiaries.

Chapters 11 and 12 widen the approach to considering the situation when the project phase has been completed, in other words, when the capital has been invested and the facilities created. Two aspects of this wider approach can be discerned. The first of them is the main concern of management in the post-project phase, covering the management of commissioning, and then the subsequent management of the operation and maintenance of the assets in a viable and productive condition. It is worth bearing in mind that a common comment on development initiatives at the present time is that they 'fail' because of poor standards of operation and maintenance, rather than because of problems encountered during the project development phase, though the latter are no doubt real enough.

The second element of this chapter is a review of the management of the agencies and enterprises which implement projects and subsequently operate them. Projects, and the assets created by them, do not exist in isolation but are part of the continuing survival or growth of the organizations that own them. It is therefore important to have a general awareness of the main concepts of management of those organizations, and how these may affect, or be affected by, detailed project management.

The reader will be aware that project management in itself is a large subject and it has not been possible to devote sufficient space to many aspects of it. This constraint is even more acute in this chapter. Indeed, it is not intended to be comprehensive in its coverage, but rather to give project managers a general understanding of the concepts and techniques of other types of management, and, in particular, an appreciation of how their own tasks and concerns fit into the wider picture.

Project commissioning: managing the transition

In reviewing project development it is possible, in broad terms, to distinguish three important sets of activities: pre-implementation activities (planning, financing, authorization, legal matters); implementation activities (particularly construction for capital-intensive projects, but also including such elements as research trials, pilot schemes, training and recruitment for agricultural and social projects) and commissioning. Commissioning is the stage that links project implementation to project operation, when the facilities that have been created (through the incurring of costs) are put to work to yield a stream of benefits. Commissioning, or a more detailed task within it, is likely to form the last activity in the overall project plan schedule.

The discussion which follows is based on approaches which have been used particularly for the process industries (such as oil and chemical plants) because in these the need for efficient and effective commissioning is very clear. However, the discussion has been much simplified in order to make it more generally applicable. Readers who are concerned with commissioning of process plant are referred to specialized texts for a more detailed treatment (Raine et al 1981). In other sectors commissioning has often not been given the importance which it deserves, because it was mistakenly thought that the facilities, once created, could immediately be put into operation and that no 'running in' period was required. Even in the non-capital sectors such as agricultural and social projects, this is unlikely to be true and all types of projects can usefully borrow some of the concepts used in the process industries.

Another reason for considering commissioning as a separate activity is that it is often overlooked in project planning and management. This is surprising in view of its importance as a key step to successful project operation – which is, after all, the ultimate objective of project development. It is moreover a complex and difficult process and for that reason can be expensive. Some estimates put the cost of commissioning in some process industries as high as 15 per cent of the total project cost. There is therefore all the more reason to take it into account and make a realistic plan for it but this is seldom done, perhaps because it lies comparatively far ahead at the time of project appraisal.

Commissioning may have a number of objectives, depending on the type of project, the owner's institutional framework and other factors. For all projects, commissioning will have operational objectives, involving the putting of the newly created facilities into use, finding the most efficient or profitable mode of operating them, and then training the operators or users in economic and effective operation. This will also be a period when clients or customers can be made aware of the products or services which the facilities can provide them. For mechanical and physical equipment, commissioning will also have a number of safety objectives, including passing tests and safety checks, training and testing operators in emergency procedures, and the search for hazards to which the plant or facilities might give rise. Finally, for facilities created under contract, commissioning will have contractual objectives, proving predictions of performance, passing of acceptance tests and provision of a trigger for payment stages. In addition, various other activities may be required to support commissioning. These include the recruitment and training of skilled and unskilled staff, the arrangement of technical assistance from the supplier or other appropriate sources, the connection of utilities, procurement of sufficient stocks of raw materials, and marketing or delivery of outputs. Records and documents relating to commissioning will also need to be kept. The commissioning process needs to be planned to take into account these various objectives, and the amount of resources and time required for them should not be underestimated.

It is probably desirable to draw up a separate commissioning schedule as the time for installation and commissioning approaches. In certain situations it may even be necessary to carry out a detailed critical path analysis for all the activities involved. It should be remembered that, at this stage of implementation, time may well have superseded money as the critical factor.

The point has already been made that it may be necessary to recruit and train staff before or during the commissioning process. Owners should look on commissioning as an important opportunity for their employees to become familiar with the operation of the assets or facilities. The actual responsibility for commissioning should be decided and defined in the contract and will depend on the type and complexity of the plant and the owner's capabilities. Even where the contractor remains responsible, it may be desirable for the owner to appoint a person to be specially responsible for the owner's inputs to the process. This should preferably be someone who will subsequently be involved with operating the plant, rather than someone, such as the project manager, who is concerned with implementing the project.

While much of this discussion has its origins in the process industries and finds its most obvious application there, many of the concepts can usefully be applied across all sectors. Perhaps the most significant point to be made is that commissioning is itself an important process and that it will not happen effectively without prior planning. The need for some form of commissioning is obvious for all projects which have an element of physical construction in them (for instance, an irrigation project, while primarily intended to provide a basis for people – farmers – to intensify their agricultural production, also has a major construction element which must be put to work in the most effective manner possible, if the farmers are to gain the full benefits). However, even projects in the social and rural sectors which contain very little physical infrastructure cannot be expected to yield their full benefits immediately the implementation or establishment phase is completed, since relationships with the clients, customers and beneficiaries need to be worked out and methods of functioning established and understood. Many problems occur during the commissioning stage and many assets and facilities fail to reach the planned level of benefits because of these problems. Owners and their managers need to give full weight to a consideration of the commissioning needs of their projects and to make detailed plans for them if necessary.

Managing operations

Once the commissioning process has been satisfactorily completed, the assets and systems created by the project can begin to operate in their intended manner. These operations will, however, themselves need management if they are to be successful. Operational management has three major features (Figure 11.1): procedures, institutional development, and resource control.

Readers may like to contrast these with the functions of project management (highlighted in Figure 3.3) and to reflect that, while there are many points of commonality, important differences relate to the acquisition and control of resources (both human and material) and to the fact that operational management is a continuing ongoing activity, unlike project management which is one-off and time-bound.

The first aspect of operational management is concerned simply with the operational procedures of the project, covering such matters as what has to be done, who has to do it when and how it is to be done. Records and safety matters are also an important consideration, as is maintenance. The latter is such an important subject that the requirements of maintenance managements are dealt with separately.

Institutional or human resource development is concerned with the management of the people within the organization, without whom it cannot continue to survive. Many of the aspects of organizations which one considers in relation to time-bound project development (ie the implementation phase) are only partly relevant when considering an ongoing, operational organization. Again, this is such a major aspect that the requirements of human resource management are discussed separately.

Resource control is concerned with the management of capital assets such as money, materials and machines in order to assist in the efficient operation of the project. In development projects the control of finance for operations through recurrent budgeting is a particular problem which

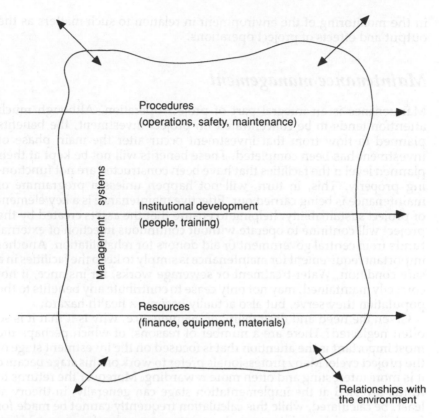

Fig. 11.1 Managing operations

is receiving increasing attention. One of the common weaknesses of project development is that there is often no direct link between the generation of benefits through the operation of assets and the provision of recurrent funds for operational costs. This may be particularly the case in agricultural and social projects, where recurrent funds are obtained from treasury subventions, which are especially prone to arbitrary and damaging reductions at times of economic stringency. Control of physical resources such as material and machinery requires systematic stock control, as briefly discussed in Chapter 8. This may also present particular problems in development projects, as it involves a whole range of disciplines such as financing, procurement, records, accountability and the like, which may be subject to cumbersome bureaucratic procedures or unavoidable external constraints. Most of the standard texts on these disciplines from industrialized countries are of little relevance in this respect.

Relationships with the environment are as important during operations as during implementation, indeed perhaps more so, as the facilities exist to sell products or provide services to potential clients within the environment. Management systems, such as monitoring and control systems, have an important role to play both in internal management and in the monitoring of the environment in relation to such matters as the output and effects of project operations.

Maintenance management

Maintenance is an integral part of project operation. Although much attention tends to be concentrated on project investment, the benefits planned to flow from that investment occur after the main phase of investment has been completed. These benefits will not be kept at their planned level if the facilities that have been constructed are not functioning properly. This, in turn, will not happen unless a programme of maintenance is being carried out. Effective maintenance is a key element of project sustainability, helping to ensure that the assets created by the project will continue to operate without continuous injection of external funds from central government or aid donors for rehabilitation. Another important requirement for maintenance is simply to keep the facilities in a safe condition. Water treatment or sewerage works, for instance, if not correctly maintained, may not only cease to contribute any benefits to the population they serve, but also actually become a health hazard.

Given the need and desirability of maintenance, why is it that it is so often neglected? There are a number of reasons, of which perhaps the most important is the attention that is focused on the investment stage of the project cycle. Many professionals prefer to work on this stage because it is more interesting and often more rewarding. Moreover the returns to funds invested at the implementation stage can generally, in theory at least, be calculated, while this calculation frequently cannot be made for funds invested in maintenance. Is maintenance value for money? Very often, it is hard to tell. Other reasons for neglecting maintenance are that the facilities simply cannot be maintained, because they were not properly designed or constructed in the first place, or because the equipment or spare parts necessary for maintenance are not available. Finally maintenance demands a high level of managerial competence, which may not always be available. It is relatively labour-intensive and involves complex relationships between large numbers of small jobs, requiring a high level of skill in management control techniques such as work programmes and budgets.

While the main focus of attention here is on the installation of efficient maintenance systems, much can be done at the planning stage to improve maintenance procedures during operation. First, it is important to make early decisions regarding the organizational responsibility for maintenance, and to secure sources of recurrent funding for it. Efforts to improve maintenance should be incorporated in the design, covering such matters as operability, reliability, accessibility and maintainability. Maintenance should be further emphasized during subsequent stages of the project cycle, for instance by specifying 'product support' at the procurement stage and by making sure that sufficient attention is given to the development of satisfactory maintenance procedures during commissioning. Finally, as pointed out earlier, maintenance is very demanding of skilled personnel, both at management and operative level. It will there-fore be necessary to arrange appropriate training, both initially and at regular intervals thereafter, in order to ensure that sufficient trained personnel is available for efficient maintenance.

Different philosophies towards maintenance can be adopted, as shown in Figure 11.2. As with discussion of commissioning, this figure borrows heavily from the approach to maintenance required for items of mechani-cal equipment but it can be usefully applied to projects across a wide range of sectors. The objective of an effective maintenance programme is to move, as far as possible, from the crises of unplanned maintenance to the predictability of planned, preventive maintenance which minimizes down-time and which allows operators and customers to plan their work to take account of periods when the facilities are out of service. The key elements of the maintenance system which must be installed if effective planned maintenance is to be achieved are discussed below.

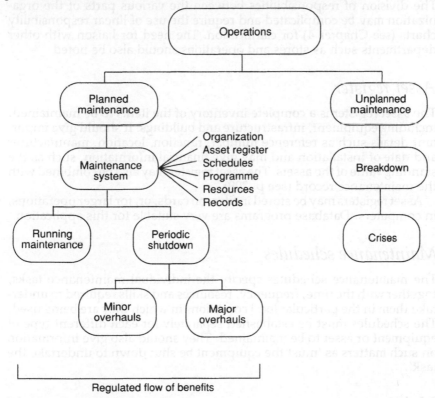

Fig. 11.2   Maintenance management

Maintenance organizations

Maintenance is a personnel-intensive activity so that organization and personnel planning is an important element. The first matter to be decided is the organizational responsibility for maintenance. Maintenance organizations may be decentralized, in which regions, districts or operating units have their own responsibility for maintenance, or 'functionalized', in which there is a central maintenance division responsible for the entire maintenance programme, with a chief maintenance engineer. For large operations there is also likely to be a need for a headquarters maintenance section, with planning and budgeting, operations, and training units, and a central workshop. There may in addition be a separate equipment organization responsible for equipment and spare parts (determining their requirements, procurement and management). The division of responsibilities between the various parts of the organization may be complicated and require the use of linear responsibility charts (see Chapter 4) for clarification. The need for liaison with other departments such as stores and operations should also be noted.

Asset register

The asset register is a complete inventory of the items to be maintained, including equipment, infrastructure and buildings. It should give important details such as reference number, function, location, manufacturer and date of installation and may give financial information, such as the standing value of the assets. The asset register may also be combined with the maintenance record (see p. 209).

Asset registers may be stored in files, on cards, or, for larger operations, in computers. Database programs are very suitable for this application.

Maintenance schedules

The maintenance schedules specify the individual maintenance tasks, together with the time, frequency, resources and skills required to undertake them in the particular local conditions in which they are being used. The schedules must be established separately for each different type of equipment or asset to be maintained. They should also give information on such matters as 'must the equipment be shut down to undertake the task'.

Maintenance programmes

The purpose of the maintenance programme is to trigger the necessary maintenance activities at the correct time, while at the same time making the most economical use of resources. The trigger may be a breakdown, inspection or, on a more systematic basis, a specified period or number of running hours. Maintenance programmes must be prepared in liaison with operations and stores staff. They lead to monthly or annual work programmes for the maintenance crews, together with budgets which define the resources required and expected costs. Much of the skill of maintenance management lies in this activity, since, even for a small operation, a large number of tasks, crews and resources may be involved.

Resource requirements

The resources which may need to be considered in relation to the maintenance programme include personnel, finance, basic equipment such as hand tools and safety equipment, stores and spare parts, workshops, and transport. Control and allocation of such resources is, of course, likely to be a major part of any manager's job. As with the establishment of the maintenance programme described above, special levels of skill in resource allocation may be required for maintenance management.

Maintenance records

These are records of the maintenance carried out on a particular item, specifying the work done, when and by whom. A cost control system is also required, to ensure adequate control and apportionment of costs in the maintenance department. From these controls, information may be derived on the performance of equipment (eg percentage of downtime), indicators of reliability, consumption of spare parts and costs. These may in turn lead to a revision of maintenance programmes and perhaps of procurement strategy.

Institutional development

The management of procedures and resources have previously been noted as two of the key aspects of operational management. Resources include finance, materials, equipment and the like, but human resources, the people who staff the organization that implements the project and subsequently operates the facilities, are separated into a special category. The people in the organization are such an important component that their management needs to be considered separately, under the general heading of institutional development.

The concerns of managing people in an operational organization are different from those in a project organization, for reasons which were discussed at the start of this chapter. Operational and agency management is a long-term continuing process, with a general goal of survival and growth. The framework of managing people in this situation is therefore conditioned by this perspective and borrows from many of the ideas of staff and personnel management, which are often not directly relevant in a project situation. Nevertheless certain 'people' management skills are common across both projects and operations, including, for instance, the skills of motivation and leadership, communication, and managing change and conflict. Project work may be a powerful motivating factor by itself and many of those involved gain considerable satisfaction from achieving project tasks or goals, often in unpleasant or difficult conditions which would otherwise be very demotivating. Managers need to be aware that the intrinsic satisfaction of achieving goals may not be so readily available in operational management, and other means for creating job satisfaction and motivating staff must be looked for.

These other means must, in general, work from the premise that the people in the organization are a long-term asset with long-term developmental goals of their own. Indeed human resource management is often called human resource 'development' to highlight these goals. A number of concepts are significant in this development, some related to the work and others to the workers themselves. Work design and job analysis is important as a method of finding ways to make the work more satisfying and challenging. Operations, unlike projects, are often repetitive or cyclical, bringing with them the dangers of monotony and boredom which in turn lead to dissatisfaction and lack of quality. As far as the workers themselves are concerned, important concerns of the manager are connected with recruitment and selection: in a dynamic economy there may be a considerable staff turnover, generating a need for replacement. In any case any vigorous organization will want to be employing a certain number of younger staff to ensure continued vitality. Once staff have been employed there will be an ongoing need for appraisal, in some form or other, to maintain satisfactory performance, and a corresponding system of rewards and incentives (pay, bonuses, and rewards in kind). Throughout, there must be a general intention to assist staff in their own long-term development, throughout their career.

Training is, of course, an integral part of any staff development plan and increasing importance is being given to it. Broadly speaking training may be divided into two types: on-the-job training and training away from the workplace. Within these two broad categories there are many further distinctions. For instance on-the-job training has traditionally meant working closely in a job and being supervised by someone already skilled in doing it. Nowadays more systematic on-the-job training systems have been devised to increase the effectiveness of this approach. Off-the-job training can also take a variety of forms, the most important distinction being between training which is formal in its approach, based on classroom techniques, and non-formal training involving discussion, case studies, role-play and other methods. Some approaches, such as action learning, have elements of both on-the-job and off-the-job training within them.

Organizational development (OD) techniques are a collection of techniques and approaches which make use of many of these concepts and others (such as the use of change agents) to develop organizational effectiveness. Though developed initially in industrialized countries they have some applicability also in the situation of developing countries. For instance many of the techniques were used in the transformation of the National Irrigation Administration of the Philippines from a typical public-sector bureaucracy into a more responsible and flexible organization serving its customers' needs (see Chapter 4). It is, however, worth repeating that such techniques are more applicable to long-term operation and institutional development, rather than short-term and task-orientated project activities.

The management of public agencies and enterprises

The main part of this book has been concerned with the concepts, techniques and skills of project management while the first part of this chapter has reviewed the key requirements for managing the commissioning and subsequent operation of the project. The final part of this chapter moves the focus beyond projects to look at the management of public agencies and enterprises. Government intervention in economic and social activity is a well-established feature of developing countries and commonly results in the creation of publicly owned organizations providing a wide range of goods and services. Throughout this discussion such organizations will be referred to simply as public agencies, taking this to apply both to organizations which provide a non-profit service to their clients or customers, and enterprises which sell goods in the market. The former group are characterized by the public utilities such as power, public health and transport undertakings, and also by many rural and social institutions such as irrigation agencies and agricultural trading boards. Public enterprises will commonly be found in the commercial and industrial sector, in mining, banking and tourism. For example, industries such as steel manufacturing are often within the public enterprise system in developing countries.

Projects are the mechanism by which the operating assets or facilities of agencies are created. These assets then continue in operation, yielding a stream of benefits which may be financial, economic or social. A typical public agency will have a set of objectives, some of which are formally established for it by government but others of which are created within the agency itself and may not indeed be formally defined. The agency will use projects as a way of creating assets which help it to achieve those objectives in the long term. At any one time, the agency will be operating some assets or systems, creating new assets through projects, and at the same time retiring or rehabilitating older assets which no longer effectively fulfil the function for which they were designed. It is worth contrasting this view of projects with that which tends to see them in isolation, as the only vehicle of development – a view which has been supported by the dominance of international lending agencies in development, and their preference for project or programme lending. This approach has led, in some cases, to the establishment of completely new organizations to implement major projects or programmes. Examples of this are the creation of separate ministries specifically for the implementation of integrated rural development programmes (even when such ministries thereby impinge on the responsibilities of other ministries already in existence, as was discussed in Chapter 1 with the case of Sierra Leone).

Another example is the establishment of the Mahaweli Authority of Sri Lanka (see Chapter 4) to implement the major projects within the Mahaweli Irrigation Programme even when there was already an existing Irrigation Department charged with the general responsibility for developing irrigation facilities within the country.

Those working in public agencies in developing countries are therefore concerned with different types of management. There is, first, project management, which has been the main focus of our interest so far. By project management, we mean the management of the creation of the assets, and in particular the implementation, establishment or construction phases. Second, assets, once created, require to be operated, generating a need for operational management. This includes management of the day-to-day operating procedures of the facility, management of resources (financial and material) and development of the human capital of the system, all on a continuing basis. Unlike project management, which is concerned with the time period of the investment, operational management works on repeating cycles, which may be weekly, monthly or yearly, depending on the management level. As can be seen, the distinction between project management and operational management is not as sharply defined for projects in the rural and social sectors, where the establishment and operational phases may overlap significantly, as in capital-intensive projects where completion of the investment phase is marked by clearly defined milestones. Third, there is a need for the management of the public enterprise or agency itself. Here managers become involved with a much wider set of concerns, of strategic, long-term planning and management, involving constant redefinition of the agency's objectives and 'mission', of the need to change as development takes place and the environment changes, and of appropriate actions and practices to ensure survival and growth.

The aspects of project management, operational management and agency management are a feature of the private sector as much as the public sector. Public agencies, however, have a policy role in addition to, and often distinct from, their organizational role, which makes for important differences from the private sector and sometimes complicates their management considerably. As an example of this, public agencies are often established to provide a mechanism for the acquisition and development of particular skills within a country, in a way not required of private enterprise. Indeed public agencies are commonly different in structure, size and technology from their counterparts in the private sector so that their management framework differs considerably. Sometimes public agencies face basic inconsistencies in their stated objectives: for instance, supply or trading corporations in agricultural areas may have been established to improve services to poor farmers, as well as to trade profitability when in fact it is impossible to do both at the same time. Often the negotiations and decisions of public agency management belong in the public domain and are subject to political interventions and public comment. Managers of public agencies thus face a special set of problems which are exacerbated by the turbulent and dynamic environment in which they exist, and by the poor resource base from which they must operate. The political dimensions to management are common to public agencies in industrialized as much as in developing countries but the problems of a turbulent environment and a poor resource base are particularly severe in developing countries.

The general framework of the management of public agencies is shown in Figure 11.3, which also shows the relationship between agency, operational and project management just discussed. While the space available in this book does not permit a full treatment of the subject, it is worth commenting briefly on the key aspects of strategic management, the institutional framework, internal management processes, and performance indicators.

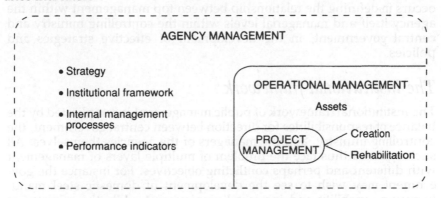

Fig. 11.3 Agency management

Strategic management

Projects are closed, time-bound undertakings, while public agencies are open, continuing systems. A key difference between them is the need for strategic, long-term management of the agency to ensure its survival and growth, and a good deal of attention is now being given to effective methods of strategic management. Such strategy must be based on an assessment of the agency's values and objectives (what is it trying to achieve, and what mix of policies is it following in trying to achieve them), its human, financial and material resources, and the environment within which it operates (how dynamic and turbulent it is, and how favourable or hostile to its activities). Strategic management of public agencies is complicated by the nature of the goods and services which they are expected to provide, ineffective management, and political interference. A key element of strategy lies, of course, in satisfactory financial performance over the long term. While all public enterprises have many other objectives, good financial performance generates further profits and increases the degree of autonomy enjoyed by the enterprise.

Public agencies are similar to their counterparts in the private sector in that the role of top management relates primarily to strategy for the organization and its relationship with the environment. The job of these

213

managers is to take a long-term view of the nature and function of the organization and its dealings with its customers, clients and suppliers so as to establish an overall policy for survival and growth. A major area of action for top management concerns the translation of high-level, long-term policies for the organization into executive and operational tasks, the monitoring of performance, the exercise of direction and control where necessary, and the facilitation of the development of an appropriate set of operating procedures. In this respect it is helpful to associate top management with decisions relating to major initial investments and to try to avoid, as far as possible, rapid turnover of staff so that responsibilities for previous decisions can be laid elsewhere. Another problem occurs in defining the relationship between top management within the agency itself and managerial levels within the controlling ministry and central government, in the development of effective strategies and policies.

The institutional framework

The institutional framework of public management is complicated by the balance of responsibilities for direction between central government, the controlling ministry, and the managers of the agencies themselves. All such agencies thus face the problem of multiple layers of management with different and perhaps conflicting objectives. For instance the government may wish to see the development of domestic steel manufacturing capability and increased employment while the ministry of industries wishes to operate common managerial and accounting procedures across all the enterprises for which it is responsible and the managers of the steel mill themselves wish to trade profitably and secure financial and managerial autonomy.

Various measures can be taken to develop a suitable institutional framework within which public agencies can operate. One measure is to improve co-ordination and planning at the national level, so that agencies have a clear set of policies and objectives defined in detailed corporate plans, drawn up during consultations between agency management and government. A mechanism for setting and agreeing objectives is to distinguish and clarify them by drawing up a contract between the agency and government. Such contracts define the respective responsibilities and obligations of the two parties and have been used, for instance, in France and some of the francophone countries of Africa.

Once a clear set of objectives have been defined there still remains the problem of how to delegate operational control to the agency management in order to provide maximum autonomy. Governments are frequently reluctant to institutionalize delegation of powers and authority to a public agency, since they see the agencies as an arm of government policy and want to retain the power to intervene. Some central control is indeed necessary, as long as the agencies remain within the public domain, but efforts should be made to restrict the level of operational control and the capacity for interference, since this reduces the manager's ability to manage and lowers morale.

Internal management processes

While the institutional and organizational framework underlies the whole basis of the agency's operations, a second key aspect to be considered is its internal management processes. Common problems encountered in the internal management of public agencies relate to a scarcity of managerial talent and expertise, which is exacerbated by low pay and high turnover as managers are moved from one agency to the next with government and political changes, and to the nature of public agencies themselves and their relationship with government, resulting in government interference, conflicting objectives, unclear decision processes and lack of managerial autonomy and morale. The two groups of problems are, of course, linked and interconnected at many points and mutually reinforce one another. For instance, government interference often increases when managers are seen to be inefficient, thus reducing their autonomy and feeling of responsibility. The management of public agencies is a specialized discipline, needing specific personnel planning and the development of distinctive skills in handling political, bureaucratic and administrative relationships, as well as the more traditional managerial skills.

Management and control at the agency level require a number of managerial approaches and actions which are found across the whole range of managerial situations. For instance, there needs to be corporate planning at appropriate frequencies, with detailed task and activity planning increasing in frequency at lower managerial levels within the organization. Agency management will need to establish action plans, to achieve the targets and policies set it by government, and to monitor the performance of the action plans on a regular basis. Management will also need to establish policies relating to the development of the human resources of the agency, and the training necessary to achieve objectives at various levels. An efficient management information system is required, to provide accurate and timely information to the managers to guide their actions. The information system is of course partly based on the definition and measurement of suitable performance indicators, as we shall discuss in the next section.

Performance indicators

Performance indicators are necessary to facilitate control and management of the agency in its day-to-day operations, and to provide the basis for strategic long-term direction. However, perhaps the greatest contribution that good performance indicators can make is in fostering a positive institutional attitude within the agency. In the absence of indicators there is a tendency to manage through formalized bureaucratic procedures. Performance indicators allow the agency to substitute output orientation for rule orientation.

215

There are potentially a large number of these indices available, covering such matters as general performance, management, financial, investment and physical performance. Selection of appropriate indicators will depend on how well they satisfy the important criteria of the degree to which they are quantifiable, communicable, valid and cheap. The use of financial statements to determine commercial and economic performance, an obvious part of any management system, is not straightforward in this case because many agencies and enterprises are monopoly suppliers, so that prices are not subject to market forces. In addition they may have preferential access to subsidized inputs and finance, so that financial statements present a distortion of the actual economic performance. A good deal of work is being done on methods of improving the measurement of economic performance of public agencies: readers are referred to specialized texts for a fuller treatment of this subject. In any case, as interest continues to be focused on public agency management, it has become clear that attention must be paid to management indicators, such as capacity utilization and productivity, as much as to financial indicators. As with commercial and economic performance, it is necessary to keep in view that performance of public agencies may be affected by their need to pursue multiple and sometimes conflicting goals.

References and further reading

Cockerill H 1987 Maintenance considerations for new plant. *International Journal of Project Management* **5 (2)**.

Kerrigan J E, J S Luke 1989 *Management training strategies for developing countries*. Boulder, Colorado, Lynne Runner.

Klatt L A, R D Murdick, F E Schuster 1985 *Human resource management*. Columbus, Ohio, Charles E Merrill.

Powell V 1987 *Improving public enterprise performance: concepts and techniques*. ILO Management Development Series No. 22 Geneva, ILO.

Raine H P, M Walsh, S H Wearne 1981 *Management of the commissioning of industrial projects*. School of Technological Management, University of Bradford.

World Bank 1981 *The road maintenance problem and international assistance*. Washington DC, World Bank.

CHAPTER 12

Current issues in development management

Carolyne Dennis

Introduction

Current concepts and practice in development management derive from a wide variety of sources: the influence of trends in management theory and the priorities of development agencies at a given time and the interpretation of the lessons to be learned from the experience of managing development programmes and projects are some of the most important influences. The diversity of this list shows the wide range of influences on development management practice. It also suggests that 'making sense' of such a wide range of sources might be a difficult and confusing experience. At present, development project management is a very volatile subject, beset by a flood of new ideas. Recent development experience has created uncertainty over objectives and the appropriate mechanisms for achieving them. The constant process of revising ideas and practice in the light of experience has been intensified by the raising of the fundamental question of how to ensure the sustainability of development initiatives. The problem for the managers of development projects is: how to learn the valuable lessons of recent experience without being deluged by a stream of new development fashions which may invalidate their previous experience and in return provide them with little idea of how to manage development projects in the present and future.

This final chapter focuses first of all on the current unease and questioning of the practice of development project planning and management. This leads to a discussion of the way in which ideas and practice of development project management might change. It is followed by a discussion of how these shifts in direction might usefully be related to the urgent problems of development project management. Finally, the implications of a few concrete examples of recent changes in the development management consensus are analysed. These issues are related to the issue of how to achieve sustainable development. This has a special

217

significance for the practitioners of the management of development projects which are intended to confer long-term benefits but are limited to a specific time period, to ensure that the creation of those benefits can be planned and managed.

Questioning the 'development management concept'

The explanation of why the central ideas of project management change is related to the more general question of why ideas change. The related question which may or may not be articulated is: does it matter to the practical development project manager who has urgent and practical problems to solve? Ideas may change in response to fashions; everyone wishes to be up to date and contemporary and this process may appear to have very little to do with the constant problems of project management. But this understandable distrust of the abstract and ephemeral does not answer the question of why fashions in development and development management ideas appear to change so rapidly. Ideas change or new ideas gain wider currency when the existing dominant ideas do not appear to explain insistent problems which require resolution (Kuhn 1970; Hunt 1989: 1-6). Ideas are valued and usually 'taken for granted' for the solutions they propose in the problems confronted. People become concious of our 'common sense' as a system of ideas that could be changed when looking for a guide to the resolution of insistent problems which will not go away and to which current ideas do not appear to be relevant.

In discussions of changes in scientific theories, the concept of a 'paradigm' is used. This means that at any time there is a set of theories, concepts, methods and techniques which are generally accepted within a discipline or profession. This paradigm provides guidance as to the appropriate concerns and the relevant practice for particular subjects. It is a useful concept for the present discussion. First, it assumes that everyday practice is based on an often unacknowledged acceptance of general ideas about the world, and second, that ideas and practice can change and progress only if they are organized into a paradigm. The concept of a paradigm thus means that, in order to understand the assumptions of everyday practice, there is a need to examine what is happening to the paradigm on which they are based. This is especially true in situations in which there are pressures for change. If subjects appear to be increasingly relevant, it may be necessary to work towards a shift in the paradigm in order to incorporate them into development management practice.

It might be helpful to consider first the questions asked about development management practice which can best be understood by reference to the debate around the dominant development paradigm before exploring what their implications might be for modifying management practice. Perhaps two of the most insistent problems which have concerned project management since the early 1980s have been

1. How can the benefits estimated by project planners be delivered to the potential beneficiaries or clients?
2. How can the benefits of projects be sustained beyond the life of projects?

The concept of sustainability is used in two senses here: the sustaining of a given level of natural resources and the sustaining of the necessary institutional capacity to ensure this while sustaining the level of benefits. These questions have emerged from reservations concerning the existing practice of planning and management of development projects and whether it can resolve these problems. A difficult and challenging question arises as to how far these concerns have surfaced BECAUSE of the existing practice of project planning and management and how far from other influences. A subsidiary question is how far these concerns have emerged from planning and how far from the management experience of development projects. These questions will be carried forward with the hope that the analysis will help to answer, or at least address them.

One problem in exploring this issue is that in exercising responsibility at the project level, there may be resistance to examining the wider development picture. Indeed, one of the attractions of project initiatives is that they appear to be relatively independent of the wider economic and social environment. This is, however, only true in the short term and that indepedence may be illusory even then. In the long term, the separation of projects from the conditions that created the need for them in the first place is not possible. This present pressure on project planners and managers to take on board lessons learnt in the wider development environment and from recent experience of implementing projects, comes largely from two major sources:

1. shifts in the general debate about economic development
2. the related debate about the significance of non-economic aspects of development.

Thus in order to analyse the significance of these concerns for the management of development projects, it is first necessary to look at what is happening in the wider development debate.

Recent trends in the development 'paradigm'

The impact of economic crisis and structural adjustment

By the second half of the 1970s, a divergence of development experience had emerged between sub-Saharan Africa and Latin America and south and south-east Asia. Sub-Saharan Africa, as the most 'problematic' and crisis ridden continent, may be of special relevance to a discussion of projects which can be understood as a response to a crisis. From the second half of the 1970s, many African countries were in economic crisis with balance of payments problems brought on by the rise in oil prices but intensified by structural weaknesses in their economies. This created a cycle of indebtedness from which they attempted to escape by accepting stabilization and structural adjustment loans from the IMF and World Bank respectively (Killick 1984; Mosley et al 1991). It is a matter of some debate as to whether these policies have been effective in Africa but it is certainly clear that they have led to a rapid decline in the standard of living of particular social groups, especially the urban poor and public sector salaried workers (Colcough 1991; Jamal and Weeks 1988). One objective of all structural adjustment programmes has been to reduce government expenditure and to privatize government activities where possible. They have also had an impact on activities in which governments had an important role to play such as development projects. Central governments suffered from acute shortages of foreign currency and budget shortfalls, existing investments could not be maintained and the value of salaries fell so that public employees have been increasingly preoccupied with obtaining other sources of income to provision their households.

This crisis, or series of crises, is beginning to have a tremendous influence on the relationship between development projects and the wider policy-making and planning environment. Projects were intended to override perceived deficiencies in local planning and management capacity. The economic crisis and adjustment programmes increased those deficiencies to levels at which projects could no longer compensate for these deficiencies as they affected the management of projects themselves. The project staff could no longer be protected from the problems confronting all salaried employees. Beneficiaries had wider problems than could be resolved by a 'free standing' project. This has led to critical analyses of the implications of the financial and managerial requirements of projects and the lack of human and other resources to sustain them beyond the period of donor funding.

The 'emergence' of sustainability

The increasing relevance of both these issues has led to the development of the concept of sustainability, as a means of providing criteria by which to determine the potential long term viability of such initiatives. Sustainability has become popular because it means anything the writer requires. There are two separate strands of interpretation of this idea which are intrinsically important and are especially relevant to the present discussion. First, that the economic environment and government policy provide the enabling or limiting possibilities for projects and programmes, and second, that the long-term viability of economic and non-directly productive projects depend on non-economic factors such as the natural environment, institutional capacity and gender issues. The insistence on the importance of these factors derives from the perception that existing methods of appraising projects tend to systematically exclude them and thus make it difficult for them to be included in managerial responsibilities.

As was stated above, the planning and implementation of development through projects is intended to overcome the constraints in the macro-economic and administrative environment. Experience increasingly indicates that the very factors which make projects seem an urgent necessity will, in the end, constrain the possibility of project managers achieving project objectives. The productivity and income increases expected from agricultural projects depend as much if not more on the mechanisms which determine agricultural prices and facilities for marketing as on the efficiency with which a project is managed (World Bank 1981; 1989). In a health project, the take-up rate for health facilities intended for children may be closely related to the possibilities for women's access to an income (Momsen and Townsend 1987). It may be possible for these factors to be favourably influenced in the process of planning a project but this is likely to be only at the margins. The problem then becomes how to manage projects with greatest effectiveness within these constraints.

The consideration of the impact of macro-economic developments on project management has also to include a discussion of the impact of economic crises and restructuring policies on the lives and potential of project managers. Projects tend to rely for their impact on overriding the constraints of line administration and sector management. However, project managers are subject to the same economic conditions as other members of a society and have to survive the same falls in their income and necessary investments in time and social networks to secure essential goods. Only foreign managers employed by donor organizations or foreign companies, under special conditions, are not affected by local conditions to any significant extent. But widespread use of this strategy tends to suggest that successful project management requires that managers are insulated from local economic and social pressures. This does not assist the development of projects which can be managed in a sustainable manner in line with local circumstances and pressures.

Non-economic influences on the management of projects

In addition to the problems created for project management by the changes in the macro-economic environment, there have always been a series of non-economic factors which have been put forward as affecting the success of projects. These issues have apparently changed through time and the debate tends to be focused on 'soft' projects such as small farmer agriculture, rural development, non-agricultural income generation and primary health care. The debate is often framed in a critical, analytical form from a perspective outside the consensus of those responsible for planning and managing the disbursement of development funds. It is often difficult to translate them into terms which appear relevant to project planning, let alone project management. Given that this focus on the non-economic factors of development has emerged and consolidated

221

outside the aid community, it is surprising how, in the recent past, it has begun to impinge upon the discussion and practice of this community. Returning to the concept of the paradigm put forward at the beginning of this chapter: the increasing receptivity of development professionals to these ideas indicates that they are aware of problems they define as 'significant', to which the existing consensus does not suggest an adequate resolution.

The three 'non-economic' issues appear to be most urgent at the present time, that is the natural environment, gender, and the household and institutional capacity. These are considered below and their significance for project planning and project management are discussed.

Environmental issues in project management

The natural environment has recently emerged as a key issue in the planning and management of development projects especially in agriculture, forestry and hydroelectric projects. This issue has become an important one for donor agencies mainly because of pressures from the electorates of industrialized countries. There are, however, a number of substantive issues which have emerged from this discussion: the first is that it is necessary when planning projects to plan for the period beyond the completion of the project. This raises the importance of two related issues of great relevance to project management: the need to use local knowledge and the need to build on, encourage and create institutional capacity (discussed separately below). Local knowledge is perceived as being of special importance in environmental concerns but, once the need for sustainable development is addressed, it follows that local knowledge of institutional and farming matters needs to be understood and incorporated into project management practice.

Gender factors in projects

Gender has entered the development debate through the perceived problems of raising agricultural production, especially in sub-Saharan Africa since the 1970s. In the 1980s the increasing significance of the gender dimensions of development has emerged from a combination of issues raised by the impact of macro-economic policies and a growing realization of the gender implications of project planning and management initiatives within different sectors. Some of the significance of the focus on gender is that it makes it possible to raise general development issues which are difficult to raise any other way in relation to project planning and management. The ability to assess the impact of adjustment policies through the everyday experience of people in many developing countries and to evaluate 'development policies' has been greatly assisted by disaggregating their impact on the household. This disaggregation has been made possible by the result of the gendered focus of many of the more influential participants in the discussion (Moser 1989). The other line of gender arguments derives from the experience of the impact of projects, especially in the areas of agriculture, rural development, rural and peri-urban water supplies and primary health care (Elson 1991).

The significance of 'gender' issues at the project and programme level is that they are linked to the same wider issues as macro-economic gender issues: the disaggregation of the household, the household division of responsibilities and resources and gender power relations in the household, public organizations and decision-making bodies. Once these issues are recognized as being central to the understanding of development and the implications of development policy as has been the case during the 1980s (Staudt 1991), the question arises of how it is possible for projects to achieve their development objectives without taking these issues into account. Following on from this, it can be argued that it is not possible to assess the impact of projects without taking these issues into account. The question becomes one of how is it possible for these issues to be incorporated into the planning and management of projects and programmes.

The relationship between institutional issues and the management of projects

Both the issue of the natural environment and of gender address different aspects of 'sustainability', but there is a further issue of sustainability which is especially relevant to activities planned and managed as projects; that of institutional sustainability. Projects are intended to be limited in time and space. It is argued above that the seductiveness of the project format to donors and governments is that it appears to make it possible to bypass the administrative, managerial and other resource constraints that make development initiatives so difficult within existing national and local administrative frameworks. Experience, especially with complex, multi-sectoral rural development projects suggests that, in the longer term the institutional incapacity of sectoral departments and local administrative structures cannot be avoided by establishing projects (Birgegard 1988). They absorb scarce administrative and management resources. In the long term, benefits can be sustained only by there being an institutional framework to which to hand over.

For all these issues – environment, gender, institutional development – it has been suggested that these have implications for the planning and management of projects. No attempt has been made so far to distinguish planning from the management issues that they raise. This is partly because these issues cannot be separated. The planning and design of projects establish the objectives and targets for management. If the discussion is widened to include the paradigm utilized earlier, the planning methods and techniques establish the framework within which management takes place and the imperatives to which management responds. This is supplemented by the practical skills of project managers in 'making things work'. These skills are particularly important in a situation in which planning priorities and methods are shifting.

223

Much of the debate around such issues as the environment, gender and institutional development are part of a long-term debate about the place of non-economic issues in development. These issues have perhaps become more important in the 1990s as the neo-classical development paradigm has broken down. The significance of these non-economic issues is that they have implications for policy and planning, the assumption is that, if these issues were taken into account, they would alter policy, planning and management, and thus the outcome of development initiatives. The process of transferring an awareness of the significance of these issues for development to the emergence of planning and management practice which incorporates them is extremely problematic. It has never been easy to operationalize these issues in a way which makes it possible for planners to use them to take decisions without losing the insights which make them valuable. It requires the confidence to combine quantitative and qualitative techniques and to move away from reliance on a few financial and economic appraisal techniques. This usually means being prepared to accommodate higher levels of uncertaintly in planning and being satisfied with lower levels of projected output and more modest targets. This has important implications for the manner in which donor development agencies operate and take decisions (Commonwealth Secretariat 1989).

Incorporating these issues into project management practice

For the purposes of this discussion, it is important to distinguish between two related levels of management: first, the conscious implementation of the targets and projected output of project plans, and second, the methods employed by project managers to make projects 'work'. When there are clear, seemingly permanent characteristics of project plans then the everyday management strategies and techniques of managers may be the source of creativity and discovery which acknowledges non-financial and non-economic factors in development. On the other hand, when project planning and design is shifting and changing, as is happening at present in the major development agencies, it is not necessarily true that management practice will reflect this shift of emphasis. It is therefore necessary to examine particular cases to see what is the basis of the management paradigm and management practice. One step towards this is the systematic examination of the implications for project management of the incorporation of issues such as environmental sustainability, gender and institutional development.

A concern with the natural environment and a preoccupation with finding a form of development which will not degrade a limited stock of natural resources became increasingly important in the late 1980s. This trend in development has been informed from the outset by an awareness of the negative environmental impact of existing development projects, especially in the agricultural, hydroelectricity and forestry sectors (Conroy and Litvinoff 1988: xi–xiv). There has recently been a number of attempts to introduce environmental impact analysis into project planning. This debate has often centred around the attempt to quantify and give a financial and economic value to environmental impact. It is still not clear how far this is possible while retaining the most important issues concerned with environmental sustainability.

In planning, a concern to incorporate environmental issues raises at least the following questions. Environmental knowledge relevant to particular development initiatives is not held in organizations such as the World Bank, Ministries of Finance and Economic Planning or even in Ministries of the Environment. One may doubt that knowledge of economic development is concentrated in this way. It is certain that knowledge of the environment and the impact on it of particular forms of agricultural development, exploitation of water resources or priorities of forestry programmes will be local, specific, probably unsystematic and hidden from conventional bureaucratic and planning processes. The first problem is to recognize the ignorance of planners and the significance of this 'hidden' local knowledge. The second problem is to retrieve it in a form relevant to development initiatives. A concern with environmental sustainability suggests that plans must contain provision for appropriate management and maintenance of development initiatives. For many rural and peri-urban projects, this means that the local population must be involved at least to the extent that they take responsibility for this environmentally sustainable management. The planning, and subsequently the crucial management issue, is how to obtain the local 'community' commitment to this responsibility.

It is very difficult to incorporate these 'environmental' issues into project plans except at a rhetorical level as they imply the downgrading of the very characteristics which make projects so attractive to donors and planners: that they have specific, preferably optimistic targets, and they appear to make it possible to override the perceived deficiencies of the local administrative and management environment. If these issues are not incorporated into the plans and designs of projects, they are unlikely to become matters of vital concern to project management. However, project managers also depend on their knowledge of what is possible in particular locations and under specific circumstances. This can become a more systematic awareness of the necessary preconditions for achieving project objectives. These are likely to include local knowledge relevant to the project and also, increasingly, the realization that maintenance and management of facilities are crucial for environmental sustainability and have to be the responsibility of local organizations and communities. These issues of project sustainability have emerged from the experience of project managers. The problem now is to incorporate them into project plans so that they become a legitimate, central concern of project managers.

The background to the emergence of gender and the disaggregation of the household has been discussed above. Two key influences have been the effects of the economic crises and adjustment programmes of the

1980s and the reports of project managers and the conclusions of project evaluation reports, especially for agriculture, rural development and health. Again, it has proved difficult to move from intellectual recognition of the importance of this issue to incorporating it into the planning procedures for projects. It requires the development of appropriate methods of data collection to make women's work more visible, and to illuminate differences within households in terms of members' access to resources and the division of responsibilities. It also requires that these issues are incorporated into monitoring and evaluation procedures. It requires that women are allocated managerial and maintenance recognition, and resources and training for responsibilities which they are likely to undertake anyway. This process would also be assisted by improving the access of women to planning and management training, and positions at local and national level in these sectors. It will be noticed that these planning issues are very similar to those identified by the discussion on environmental planning issues.

The same general comment applies to the incorporation of gender issues into project management as for environmental sustainability. It is difficult for project managers to make them a priority if they are not incorporated into project designs with the project targets and objectives being modified accordingly. But it is also true that one of the reasons that gender has become an issue for planners and managers is that they have found it is difficult to manage their projects without taking women's work and responsibilities into account. Perhaps the way forward for managers is on two different levels of generality. There may be limited gains from focusing on women's responsibilities. One result of this has probably been an increase in women's work by taking on project responsibilities (Palmer 1991). There may be a need to go on to consider gender relations and their relevance to project design rather than focusing on women as a separate 'problem' or resource. At the more practical level, it implies being willing to develop methods of collecting data which are relevant to securing significant gendered data in a particular place. It also means incorporating local women, who may be perceived as passive beneficiaries of projects, as managers to ensure its sustainability and to enable them to gain access to the benefits. This then raises the question of how early these future managers have to be incorporated into project planning and management – probably earlier than is commonly the case. This would imply a willingness to set more flexible targets and objectives for projects and to perceive the process of consultation, agreed methods of data collection and institutional development as themselves being targets and outputs of the project.

The brief discussions of the natural environment and gender have both raised the question of institutional development: what kind of institutional structure is appropriate to enable these issues to be incorporated into project plans and management? How can these structures be brought into existence or strengthened and what modifications would this involve for methods of project planning and management? This issue of institutional capacity and development also has other dimensions. They have arisen largely because of repeated evidence from project evaluation reports and post-project experience that, after projects finish, it is very difficult to maintain the benefits. The inadequacies of local administration, planning and management which projects are intended to override are intensified by the use of projects which absorb scarce local management and financial resources. They also tend to 'sideline' any local government development resources and responsibilities. This makes it even less likely that they will be able to take over the responsibilities inherited from, say, a rural development project. This problem has been made worse in countries with a recent history of economic crisis and adjustment programmes which have reduced the financial and human resources available to local and national government. Local government, especially when distant from decision-making urban centres, suffer most from this process. When the question is asked – how can the benefits of projects be sustained? one major component of the answer has to be sought in the development of locally grounded institutions which will maintain and manage assets and services.

If institutional development is to be incorporated into project management, it needs to become a priority for project planning. This requires considerable modification in the priorities of planning and the expertise which is regarded as relevant. Financial resources are needed to provide for the necessary training and time to be allocated for training and practice to take place. The process of establishing or supporting institutions becomes the target of project plans and influences the form of monitoring and evaluation procedures. This development implies a 'loosening' up of the planning process with more flexible objectives, targets and programmes of work. It implies a shift of emphasis in the forms of expertise which are perceived by donor agencies as being relevant. It also means acknowledging that other organizations and agencies may need to become involved in the planning and management process, such as local government bodies and non-governmental organizations, which reduces the control of the project planners and managers over their own project. In summary, it means being willing to carry higher levels of apparent uncertainty than donor agencies' project planners have previously been willing to do.

Some of these issues are carried from planning to management: in general, the reduction in control over the project and the need to work with other organizations, the uncertainty which results from a focus on strengthening the capacity of institutions to take over from the project management and a resulting change in the appropriate 'mix' of qualities required by project managers. However, it is at the management stage of a project that these issues become most relevant. They are often threatening to project management and difficult to handle. It is not easy to retain a sense of professional competence in the light of the need to negotiate with organizations which have limited ability to adapt to completion targets linked to the capacity of community organizations to take over project responsibilities and to adopt a lower profile than is customary for managers who like to 'lead from the front'. Making oneself redundant is always more difficult in practice than in theory.

These issues of environmental sustainability, gender-conscious development strategies and institutional development have common characteristics: they imply a moving away from the clear separation of the project from its 'environment' and a willingness to be more flexible about targets, management strategies and short-term management priorities. The discussion could be regarded as partial as it is particularly related to developments in decentralized development, rural and peri-urban development, smallholder agriculture and local level health and educational initiatives and would not be so relevant to large-scale industrial or infrastructural projects. But these problematic areas will remain the priority development investment areas for the 1990s. The debate about possible shifts in planning and management practice can be seen as an attempt to develop forms of project management which are appropriate to these sectors (Oakley et al, 1991).

In conclusion, although it is risky to second guess what is going to happen in the field of project management over the next few years, it does appear that project managers are going to live in 'interesting times'. Projects retain their importance as the 'cutting edge' of development but they are required to be more flexible, able to adapt to local conditions, to modify project objectives in the light of experience and to make themselves redundant. These are extremely difficult developments to incorporate into conventional perceptions of managerial competence. There is no easy way to adapt to these developing management priorities but project managers are going to be required to have the skill and confidence to adapt to rapidly changing conditions. The one certain fact is that the 1990s are going to be a period of rapid and maybe radical change in the demands made of project managers.

References and further reading

Birgegard L E 1988 A review of experiences with integrated rural development. *Manchester Papers on Development* **IV** (1) (January): 4-27.

Colcough C 1991 Wage flexibility in sub-Saharan Africa. In G Standing, V Tokman (eds) *Towards social adjustment: labour market issues in structural adjustment*. Geneva, ILO: 211–34.

Commonwealth Secretariat 1989 *Decentralised administration in Africa: policies and planning experiences*. London, Commonwealth Secretariat Management Development Programme.

Conroy C, M Litvinoff (eds) 1988 *The greening of aid: sustainable livelihoods in practice*. London, Earthscan.

Elson D (ed.) 1991 *Male bias in the development process*. Manchester, Manchester University Press.

Hunt D 1989 *Economic theories of development: analysis of competing paradigms*. London, Harvester Wheatsheaf.

Jamal V, J Weeks 1988 The vanishing rural–urban gap in sub-Saharan Africa. *International Labour Review* **127** (3): 271–92.

Killick A (ed.) 1984 *The quest for economic stabilisation: the IMF and the Third World*. London, Gower.

Kuhn T 1970 *The structure of scientific revolutions*. Chicago, Chicago University Press.

Momsen J, J Townsend (eds) 1987 *The geography of gender in the Third World*. Albany, New York, Hutchinson.

Moser C 1989 Gender planning in the Third World: meeting practical and strategic gender needs. *World Development* **17** (11): 1799-1826.

Mosley P, J Harrigan, J Toye 1991 *Aid and power: the World Bank and policy based lending*. London, Routledge.

Oakley P et al 1991 *Projects with people*. Geneva, ILO.

Palmer I 1991 Gender, equity and economic efficiency in adjustment programmes. In H Ashfar, C Dennis (eds) *Women and adjustment policies in the Third World*. London, Macmillan: 69-86.

Staudt K 1991 *Managing development: state, society and international contexts*. London, Sage.

Toye J 1987 *Dilemmas of development: reflections on the counter-revolution in development theory and policy*. Oxford, Basil Blackwell.

World Bank 1981 *Accelerated development in sub-Saharan Africa: an agenda for action*. Washington DC, World Bank.

World Bank 1989 *From crisis to sustainability in sub-Saharan Africa*. Washington DC, World Bank.

Index

T - #0171 - 071024 - CO - 234/156/13 - PB - 9780582082236 - Gloss Lamination